ENVIRONMENTAL SANITATION AND ERADICATION OF SCAVENGING IN INDIA

By the Same Author :

Road to Freedom: A Sociological Study on the Abolition of Scavenging in India.

मुक्ति के मार्ग पर: भारत में गंदगी ढोने की प्रथा के. उन्मूलन का समाजशास्त्रीय अध्ययन (अंग्रेजी कृति "रोड टू फ्रीडम" का हिन्दी अनुवाद)

ENVIRONMENTAL SANITATION AND ERADICATION OF SCAVENGING IN INDIA

BINDESHWAR PATHAK

MOTILAL BANARSIDASS PUBLISHERS
PRIVATE LIMITED • DELHI

First Edition: Delhi : ***2015***

ISBN : 978-81-208-1380-9

Also Available at :

MOTILAL BANARSIDASS

41 U.A. Bungalow Road, Jawahar Nagar, Delhi 110007
8 Mahalaxmi Chamber, 22 Bhulabhai Desai Road, Mumbai 400026
203 Royapettah High Road, Mylapore, Chennai 600004
236, 9th Main III Block, Jayanagar, Bengaluru 560011
8 Camac Street, Kolkata 700017
Ashok Rajpath, Patna 800004
Chowk, Varanasi 221001

MLBD Cataloging-in-Publication Date

Environmental Sanitation and Eradication of Scavenging in India by Bindeshwar Pathak
Includes bibiliography
ISBN : 978-81-208-3646-4
I. Environmental II. Sanitation III. Scavenging
IV. Pathak, Bindeshwar

Printed in India

by RP Jain at NAB Printing Unit,
A-44, Naraina Industrial Area, Phase-I, New Delhi-110028
and published by JP Jain for Motilal Banarsidass Publishers (P) Ltd,
41 U.A. Bungalow Road, Jawahar Nagar, Delhi-110007

Dedicated to

Mahatma Gandhi

Father of the Nation

and

Shri Narendra Modi

Hon'ble Prime Minister of India

Contents

PREFACE

My concern and commitment to the cause of environmental sanitation and eradication of manual scavenging since early seventies, inspired me to devote my full time and energy towards upgrading the country's sanitation scenario by developing a low-cost sanitation technology. Such a technology, if and when implemented on a nation-wide scale, would result not only in the improvement of environmental sanitation, but also prove to be a more effective strategy in the elimination of manual scavenging and in liberating scavengers from the demeaning and dehumanizing practice of handling and carrying night soil, often as headload. This decision, some of my friends doubted, was 'impractical' as past efforts in this direction had made little impact. The case of Mahatma Gandhi's campaign for social uplift of scavengers was cited as an illustration. The discordant voices, however, did not deter me from following the difficult path of Gandhian struggle to do something innovative and substantial in the area of environmental sanitation, and thereby, in the eradication of manual scavenging and liberation of scavengers. My struggle had been full of many moments of despair and disappointments in the beginning; but eventually my persistence paid, and the so-called 'impractical' idea became a throbbing reality and the mission of my life.

All this happened when a small voluntary organization, *Sulabh Shauchalaya Sansthan* which I founded at Patna (Bihar) in the year 1970 became a pioneering experiment in environmental sanitation and liberation of scavengers. The experiment picked up so well in a relatively small time that it attracted the attention of the concerned governmental and international agencies in the area of environmental sanitation. The relentless efforts of our team of extremely dedicated volunteers, professionals, friends and experts in the low-cost sanitation technology – all working for Sulabh International (the new name of the old outfit – Sulabh Shauchalaya Sansthan), proved the myth wrong and demonstrated such practical results that everybody concerned with and connected to the cause of environmental sanitation and liberation of scavengers did realise that my mission of life was not merely a visionary idea. Impressive results shown by Sulabh International convinced the planners and policy-makers that Sulabh International is truly endeavouring to popularize low-cost sanitation technology, easily adaptable to Indian conditions, and, was working pretty hard on materializing Mahatma Gandhi's dream of liberating the scavengers from the age-old shackles of social disability and discrimination.

For its spectacular achievements and exemplary contribution in the field of environmental sanitation, eradication of manual scavenging and liberation, training and rehabilitation of scavengers and their wards, Sulabh International has received envious admiration and encouraging response from national and international media. Modesty does not permit me to write in detail about my achievements, but all those concerned with the subject know fully well about my commitment and contributions. Several national and international awards which I have received speak eloquently about the 'mission' I wish to complete in my lifetime. In recognition of my seminal contributions in the liberation of scavengers, social transformation, up-gradation of down-trodden, prevention of environmental pollution, creation of non-conventional sources of energy, training of change-agents for voluntary action and diffusion of innovations for promoting inexpensive and affordable indigenous technology in sanitation, health, hygiene and rural development, Padma Bhushan was conferred on me in 1991.

Here, a word about my interest in sociology is required to explain my desire to continue reading and writing about the relevance of sociology, as also pursuing research in sociology, not merely for earning academic distinctions/ degrees, but as a necessary part of my inclination and avocation as an Action Sociologist. I chose to study Sociology for understanding the society to serve it better. I loved to pursue research in 'action sociology' and earned my Ph. D. degree in Sociology from Patna University (1985) on a subject, which I had chosen for a life-time mission – The Concept of Liberation of Scavengers through Low-cost Sanitation in Bihar.

The awarding of the Ph. D. degree led me to realize that a true sociologist with deeper insight into the problems of Indian society would also like to provide a new orientation to make sociology more relevant to society. Providing action orientation to Sociology soon became my cherished desire. I discussed my idea of action sociology, initially with a small group of distinguished Indian sociologists, who supported the idea and promised to extend their full cooperation. Later, I held many parleys with senior Sociologists and received their encouraging response. In all humility I wish to put on record here that I played a key-role in establishing the Indian Association of Action Sociologists, which now enjoys enormous support of a large number of sociologists holding senior positions in the Department of Sociology throughout the country. In appreciation of the programmes and activities of the Indian Association of Action Sociologists, the International Sociological Association placed on its agenda a two-day session on action sociology during the XII World Congress of Sociology at Madrid, Spain (July 9-13, 1990).

Not content with this, I have written extensively on several important aspects of 'action sociology' in order to popularize this new sociological specialty with a view to eventually getting it accepted by the University Grants Commission. I also arranged funds for the publication of an annual journal of action sociology – *Action Sociology Review.*

I have purposely made a somewhat lengthy reference to the background information about my credentials in environmental sanitation and liberation of scavengers. I have done this to establish the fact that my interest in Sociology and social action is everlasting. While the field of environmental sanitation has been dominated by scientists and planners, I have done much thinking and writing on the subject from an altogether different angle, and that too from a sociological standpoint. I have dealt with the subject in its multiple dimensions, covering its interlinkages with the eradication of manual scavenging and liberation of scavengers. My approach has been accepted in the official, national and international circles connected with environmental sanitation and liberation of scavengers. Enthused with an overwhelmingly positive response, I have pursued the present study for over two decades.

Having studied the whole gamut of relevant national and international literature on the subject, I have been yearning for long to write and publish a book which may highlight all significant issues in regard to policy-making, planning and execution of the programme of environmental sanitation and liberation of scavengers, as also to explode many myths that cloud peoples' thinking.

After hard work of about four years, I was able to prepare the first draft of this work. In order to make it fully authentic, I had consulted quite a good number of experts, field functionaries, beneficiaries, and, of course, my academic mentors in sociology. Everyone, who glanced through the pages of the draft suggested for the submission of the material into the form of a D.Litt. dissertation of Patna University's Department of Sociology. I was initially hesitant because I did not intend preparing the draft for earning a degree. Further, I was not sure whether a D.Litt. degree would add to my eminence in the field. Since the pressure to accept the suggestion came from many distinguished sociologists, whose valuable advice has always been dear to the researcher, I concurred with the wishes of all those sociologists whom I have held in high esteem.

It attempts at a comprehensive critique of the policies and programme of environmental sanitation and liberation of scavengers. In the overall context of environmental sanitation and eradication of scavenging, it basically aims at studying the historical context in which scavenging originated and developed; examining its linkages with ecology, health, hygiene and environmental sanitation; probing into various reformative and rehabilitative measures undertaken by a good number of social reformers, voluntary action groups, Governmental and international agencies for the liberation and social upgradation of scavengers; analyzing the policies, programme and problems of low-cost sanitation schemes by Central and State governments; underscoring the pioneering efforts of Sulabh International and its outstanding achievements in the field of environmental sanitation and eradication of scavenging; describing and discussing the relevance of sanitation movement in the liberation of scavengers and the linkages of the liberation programme to the improvement in environmental sanitation; presenting an overarching view of the constraints in the implementation of policies envisioning a better environmental sanitation scenario and a brighter future for the scavengers; and finally, furnishing a sociological perspective on all important dimensions of environmental sanitation and eradication of scavenging in India.

The book is the outcome of my dissertation for D.Litt. degree in Sociology, awarded by the Patna University in the year 1994. It is a revised and updated version containing a critical appreciation of recent advances in the field of environmental sanitation in India and abroad. It embodies the author's endeavour as an action sociologist to critically look at the new dimensions and current perspectives on environmental sanitation in India.

BINDESHWAR PATHAK

ACKNOWLEDGEMENTS

The pursuit of knowledge necessarily involves learning from various sources and scholars, who enhance knowledge to perceive relevant ideas and develop insights which may lead to theoretical formulations in systematic and methodological manner, and, finally, present the social realities for sociological enquiry. In this connection I have had ample opportunities for easy access to a wide variety of these knowledge-generating sources. My thinking has, therefore, been enriched by going through relevant sociological ideas and literature. I have enriched my knowledge by the contributions of large number of men of distinction who have made a distinct mark in various academic specialties which they have carefully studied, perfected and even pioneered.

The distinguished academicians, social activists, scientists, and administrators have contributed immensely to completing several of my research publications. As a result of their continuing inspiration, I have published a book, *ROAD TO FREEDOM. A SOCIOLOGICAL STUDY ON THE ABOLITION OF SCAVENGING IN INDIA* – my Ph.D. thesis in sociology from Patna University. I was immensely influenced by several distinguished sociologists of the country. Soon after its publication, I had received many personal communications from an amazingly large number of sociologists, administrators and sanitation experts from India and abroad, praising this work as of seminal nature. I was enthused to pursue the theme on a much bigger scale with greater analytical skill. Most of my friends and well-wishers in academia, in social service and in the government persuaded me to work on a project: *ENVIRONMENTAL*

SANITATION AND ERADICATION OF SCAVENGING IN INDIA, to capture the social reality in its all possible dimensions, which, however, had remained unexplored and unexplained. As the subject is too dear to me, I had readily agreed to renew my field experiences and present them in the form of a dissertation that earned me a D.Litt. degree in Sociology from Patna University in 1994.

In course of my research work, I received immense active cooperation from several distinguished scholars and teachers, particularly from Patna University, who have contributed a great deal with preparation of my D.Litt. dissertation, as well as in the fine-tuning of the manuscript of this book.

It is against this backdrop, I deem it my duty to place on record my gratitude to all those whom I owe an enormous debt. Since the names are too many, I am remembering only those whose help and cooperation has been most valuable and most noteworthy.

At the outset, I would like to express my deep admiration for my teacher, Prof. Ziauddin Ahmad, former Head of the Department of Sociology, Patna University, whom I regard as my mentor in sociology. It is he who inspired me to pursue my research towards Ph.D. in Sociology and agreed to guide and supervise my work. I have always received tremendous academic support from Prof. Ahmad. In pursuance of the dissertation for the D.Litt. degree in Sociology he has helped me a lot and his valuable guidance was always available to me. Words are too inadequate to describe my sense of gratitude to Prof. Ahmad.

I am extremely grateful to Professor Musi Raza of the Department of Sociology, Patna University, for the kind of academic support he provided me throughout the completion of this project. I have sought his guidance on more than one occasion and have always found in him the abundance of human warmth and rich reservoir of scholarly ideas. I am deeply touched with his sense of affection for me.

I am beholden to Smt. Dr. Archana Sinha, Professor of Sociology, who has always been a great source of inspiration for me in many of my academic pursuits in sociology. In as far as the completion of this project is concerned, Prof. Sinha has extended me the fullest possible cooperation. I offer my sincere thanks for her invaluable assistance.

I am highly obliged to Prof. Phlora Sinha, Head, Department of Sociology, Patna University, for her affection to me. In fact, she has been exceptionally kind in allowing the submission of the dissertation. I hold immense reverence for her.

I must express my word of praise to friends like Dr. A.K. Lal, Dr. S.D.N. Singh and other younger sociologists of Patna University whose valuable comments on some critical issues integrally connected with the theme at hand made my task easier.

I am greatly indebted to late Shri A.K. Roy, former Chairman, Administration, Sulabh International, a senior colleague, who offered me lots of practical ideas and provided me with a great deal of research input. Being a distinguished sanitation expert himself, he had enriched my understanding of the subject in more than many ways. I feel without his cooperation and expert input, this work may not have become as authentic as I confidently find it to be.

I have special words of thanks for Prof. K.J. Nath, former Director, All India Institute of Hygiene and Public Health, Kolkata for offering valuable comments on the final draft. I am equally thankful to Dr. Indira Chakravarty, D.Sc., Professor Director and Head of the Department of Nutrition and Bio-Chemistry, All India Institute of Hygiene and Public Health, Kolkata for her cooperation and assistance in getting access to some relevant information, which may not have been available otherwise.

Since no study of this kind could have been completed without the encouragement and support of number of institutions/agencies/persons, I have naturally incurred the debt to many people who had extended their unstinted cooperation in the completion of the task. As the list of such institutions/ agencies/persons is indeed too long, I place on record my deep appreciation for all of them.

I owe special appreciation to all the distinguished action sociologists in the country for their constant advice and guidance in making a systematic presentation of facts in the approved social science style and language. I have held many informal discussions with most of them, individually, as well as, collectively. These discussions have been very valuable for me, and I have duly incorporated in this work the insights made available to me.

I gratefully acknowledge the most valuable co-operation of Prof. S.P. Srivastava, Prof. S. D. Sinha, Prof. Mangal Murty and Shri Arjun Prasad Singh in helping me to finalise the draft for publication in a book form.

Last, but not the least, I have to offer a word of thanks to all my colleagues in Sulabh International whose love and affection has always been a source of strength to me.

BINDESHWAR PATHAK

Chapter - 1
THE PROBLEM

A global quest for the preservation and protection of environment, and a subsequent quest for alternate development strategies in perfect harmony with nature is the major concern that confronts the whole of mankind today. Environmental degradation in various forms has assumed dimensions of a major crisis in both developing and developed societies. In such a global scenario, environment has occupied the centre-stage of the world's agenda for sustainable development. The problems concerning the environment had been highlighted at different fora, nationally and internationally, and serious concern has been voiced at the nature of environmental challenges that the world faces at present.

Although environmentalists have been raising hue and cry over the degradation of environment for quite some time, the global concern about the environmental issues was first seriously discussed at the United Nations Conference on Human Environment and Development (Stockholm, Sweden, June 1972). It was nearly eleven years after the Stockholm meet, the World Commission on Environment and Development (Brundtland Commission) was set up. This commission in its historic document – *Our Common Future* (1987) introduced the concept of 'Sustainable Development', and called upon all nations to address the issues of environmental sustainability, while implementing development programmes. This was followed by The Rio Earth Summit in 1992 when a far-reaching programme of sustainable development was adopted. Following the acceptance of the report of Brundtland by the United Nations General Assembly, several international institutions made fundamental commitments to sustainable development. The Rio Summit resolved to fix targets, draw time-tables and name partnerships to speedily increase people's access to clean water, sanitation, adequate shelter and food security. While addressing poverty alleviation measures, the Summit made special reference to adequate provision of water, sanitation and health care facilities to the disadvantaged people who have little or no access to basic minimum needs of life. The Rio Conference, known as Earth Summit – I, had before it a very wide and ambitious agenda on climate change, disruption of ozone layer, shrinking of tropical rain forests, loss of bio-diversities, worldwide loss of top soil and a host of other issues intimately linked to the alleviation of poverty in terms of income, health, education, food and nutrition.

The World Summit on Sustainable Development, also known as "The Earth Summit – II" and popularly referred to as Johannesburg Summit, held in 2002, reiterated the global commitment to sustainable development to ensure the relationship between the nature's resources and human needs (or greeds). It resolved to build a humane, equitable and caring global society cognizant to the need of human dignity for all. The Summit addressed several environmental issues, especially water, sanitation and hygiene. It came up with 'Johannesburg Declaration on Sanitation and Energy' to increase access to sanitation, improve human health and reduce infant and child mortality and prioritize water and sanitation in national development strategies.

The Summit recognized the centrality of water and sanitation as "essential" for sustainable development. It averred that without progress on water and sanitation front, the critical targets of sustainable development would be difficult to achieve. Water, the Summit said, is intimately linked to health, agriculture, energy and biodiversity.

Likewise, sanitation has a strong and recognizable linkage to good health and human wellbeing.

The Declaration recognized that the provision of clean drinking water and adequate sanitation is necessary to protect human health and the environment. It agreed to halve, by the year 2015, the proportion of people who are unable to reach or afford safe drinking water and who do not have access to safe sanitation. It called upon the member nations of the United Nations to initiate action at all levels to :

- Develop and implement efficient household systems.
- Improve sanitation in public institutions, especially schools.
- Promote safe hygienic practices.
- Promote education and outreach, focused on children as agents of behavioral change.
- Promote affordable, and socially and culturally acceptable sanitation technologies and practices.
- Develop innovative financing and partnership mechanisms.
- Integrate sanitation into water resources management strategies.

The Summit acknowledged that addressing water, sanitation and hygiene – the "WASH" issues – is central to the struggle for sustainable development. The Water Supply and Sanitation Collaborative Council (WSSCC) in its Vision 21 exercise endorsed: (a) the need to prioritize sanitation and hygiene both at household and institutional level; (b) the need to focus on affordable and appropriate technologies; (c) the need to ensure that infrastructure and services meet the needs of the poor and are gender-sensitive; and (d) the need for good governance, partnership and participation of all stakeholders. The WSSCC resolved to champion the Johannesburg Plan of Implementation in various ways : through the operation of its existing work programmes on advocacy, evidence - based applied research and national level action; through membership to the Water and Sanitation Millennium Development Goal Task Force; and through lobbying work at the Commission on Sustainable Development.

The Water Supply and Sanitation Collaboration Council (WSSCC) fully endorsed the Millennium Development Goals (MDGs) and averred that they offer a viable and effective framework to combat global poverty. WSSCC recognized the political importance of MDGs and sought to champion the MDG target of halving, by 2015, the proportion of people without access to safe drinking water and basic sanitation facilities. In effect, the MDGs on water and sanitation represent :

- An understanding of the intricate importance of water, sanitation and hygiene for all other fields of development;
- Willingness and commitment by all UN members to put water, sanitation and hygiene firmly on the agenda;
- A benchmark for progress and monitoring of government commitments to WASH issues.

The global concern on water sanitation has also been consistently expressed at a number of international fora. The United Nations Water Conference (Mardel Plata, Argentina, April 1977) gave priority to supplying safe drinking water and sanitation services whilst focusing on development of national water resource assessment and national policies and plans. The United Nations General Assembly (New York, November 1980) agreed upon achieving the universal access to water and sanitation through the declaration of the "International Drinking Water and Sanitation Decade (1980-1990)". The Global Consultation on Safe Water and Sanitation for the 1990s (New Delhi,

September 1990) agreed to make concerted efforts worldwide to enable people to obtain two of the most basic human needs : safe drinking water and adequate sanitation. The Second World Water Forum Meeting in Hague (2000) and the International Conference on Fresh Water in Bonn (2001) served as important holder of comprehensive dialogues to create global consensus on how to address increasing water and sanitation challenges.

These international meetings have identified several critical issues and challenges, with increasing focus on water supply and sanitation. They proposed many actions to meet the challenges, stressing the importance of taking concerted action on water supply and sanitation services as an entry point to achieve the goal of sustainable development. As a result of the sustained UN advocacy on water and sanitation, spanning about 30 years, there has emerged a very dynamic process of advancing international understanding and cooperation on key issues relating to water and sanitation. As of now, there are certain important regional and intergovernmental bodies as well as organised groups like Water Supply and Sanitation Collaborative Council, The Global Water Partnership, the Gender and Water Alliance, and the World Water Council to work for implementing the United Nation's vision of providing access to safe drinking water and adequate sanitation facilities for all.

The twelfth session of the Commission for Sustainable Development, also known as CSD-12, held in April 2004 at the United Nations Headquarters in New York, deliberated upon issues of water, sanitation and human settlements and offered unique opportunity to focus on implementing actions to achieve the internationally approved goals of water, sanitation and human settlements.

The thematic interactive discussion on sanitation revolved around the status of implementation of the Millennium Development Goals and the Johannesburg Plan of Action;

strategies for improving access to basic sanitation; and financing sanitation and reaching the poor through small entrepreneurs – employment generation activities concerning basic sanitation.

The session had identified some of the major challenges, which need to be addressed with a view to strengthening the implementation of water, sanitation and human settlements goals and targets.

- Integrating water, sanitation and human settlements and ensuring that they are inclusive, nationally driven and that their implementation is closely monitored;
- Increasing recognition of the importance of sanitation and hygiene and their linkage to health, poverty eradication and sustainable development;
- Disseminating knowledge and experience, particularly on low-cost and locally adaptable approaches and technologies;
- Building capacity for water, sanitation and human settlements planning and development in developing countries with financial and technical support from developed countries.

Environment and Sanitation

Sanitation is an integral component of environmental protection and essential ingredient of sustainable development. Indeed, sanitation and environment are two sides of the same problem; one cannot be solved without solving the other. Sanitation has many linkages, most important being water supply, personal cleanliness and safe and hygienic systems of human waste disposal. Various studies carried out in India and abroad have shown that the benefits of safe water are limited without sanitation. Sanitation ensures a productive life, the lack of which creates poverty which is the greatest polluter of environment. Putting it simply, sanitation is not only keeping clean but protecting those sources of environment which support sustainable development. The Development programmes, however innovative they may be, are not likely to yield desired results unless environmental sanitation is improved and ecology protected.

The fast deterioration of environmental and ecological balance has, therefore, become a matter of deep public concern. Population explosion, urbanization, industrialization, felling of trees, erosion of soil and an ever-increasing number of vehicles and factories are the main causes of this imbalance. A closely related issue to rapid and uncontrolled urban growth is the proliferation of slums and unauthorized colonies. Much of the World's population lives in slums, pavements and shanty towns. Their health and physical well being are jeopardized by inadequate sanitation and garbage dumped adjoining the places of habitation. The debilitating effects of insanitary living conditions and deteriorating environment lower the productive potential of the very people who can least afford it. Among the poor, the women and children suffer the most.

This backdrop of the problematics of environmental sanitation suggests that despite unprecedented advances throughout the world, the failure to extend the fundamental benefits of hygiene, basic sanitation and safe drinking water to all, remains a hurdle to development and a root cause of persistent poverty. That such a situation still exists demonstrates one of the most pervasive but silent emergencies of our time : the failure of countries to ensure safe water, basic sanitation and hygiene knowledge for all of their people.

Environmental Sanitation

With the inclusion of water and sanitation issues in the global agenda on sustainable development, specially in the Johannesburg Declaration and in the Millennium Development Goals (goal seven), sanitation has emerged as one of the major issues of environmental protection. Little wonder then that by now, there are abundant writings on almost all issues relating to sanitation. One visible effect of the ascendancy of sanitation in the global discourse on environmental protection is that the term 'environmental sanitation' (alternatively called 'ecological sanitation', or in short 'eco-sanitation') has gained wider currency in the recent writings, discussions, resolutions and recommendations having a bearing on the mandate of the Water Supply and Sanitation Collaborative Council (WSSCC).

In understanding environmental sanitation, three terms, that are frequently used in this connection, need to be understood : (i) **Basic hygiene sanitation facility :** means household latrine which safely contains human excreta, and thus protects users and the environment from adverse effects; (ii) **Sanitation :** means household sanitation including latrines, personal and household hygiene; and (iii) **Environmental Sanitation :** Covers sanitation in communities and local settlements. It also covers hygiene/cleanliness, management of wastes, including human and household wastes, and such local environmental issues like pollution of water, air and ground.

The World Health Organisation has been promoting the concept of environmental sanitation in its efforts to devise and develop suitable environmental sanitation strategies, projects and programmes, with special reference to developing countries, which confront a daunting sanitation challenge of massive proportions. The WHO Expert Committee on Environmental Sanitation defined environmental sanitation as including the control of community water supplies, excreta and waste water disposal, refuse disposal, vectors of disease, housing conditions, food supplies and handling, atmospheric conditions, and safety of the working environment. As per the WHO, environmental sanitation is a health related package which includes safe disposal of all types of waste (solid, liquid, gaseous and human), personal hygiene, safe drinking water and cleanliness inside and outside of building, domestic or otherwise. WHO considers environmental sanitation as a strategy of control over all such factors in man's physical environment which exercise or may exercise a deleterious effect on human development, health and survival.

The WHO's elaborations on environmental sanitation come closer to the conceptual underpinnings of sanitation, when it is viewed holistically. Writings on the subject use sanitation as a broad term which includes human waste, waste water, solid waste, control of vectors of disease, domestic and personal hygiene, food sanitation, housing etc. It refers to all conditions that affect health and wellbeing of the people and provide dignity and prestige to the community. Indeed, it is a science of safeguarding health.

The U.S. National Sanitation Foundation has described sanitation as a way of life and linked it with the quality of life that finds expression in clean homes, clean neighbourhoods and clean communities. A WHO Study Group defined sanitation as the means of collecting and disposing of human excreta and community liquid wastes in a hygienic way so as not to endanger the health of individuals in the community as a whole (WHO, 1987). Viewed in its composite sense, sanitation has many facets like safe disposal of human excreta, waste water disposal system, solid waste disposal, domestic and personal hygiene, food sanitation, water supply, sewerage, control of vectors of disease and prevention of environmental pollution to keep the society clean, healthy and productive. Sanitation, therefore, is not a minor issue in the promotion of the health and wellbeing of the people; it is a crucial component of social and economic development. It is an important health index of any developing country intimately linked with people's productivity and economic progress of a country.

Water and sanitation are important issues for the development discourse because diseases related to these two remain the biggest killers, especially of children. Approximately 5000 children die every day from diarrhoeal diseases due to inadequate sanitation and hygiene (UNICEF 2008). About 1 billion people worldwide are infested with intestinal worms, and as a result, suffer nutritional deficiencies and poor growth. Both these groups of diseases are transmitted through human faeces in the environment (WHO, 2003). Human excreta is the cause of many enteric diseases such as cholera, dysentery, typhoid, paratyphoid, infectious hepatitis, hookworm, diarrhoea etc. Studies reveal that over 50 infections can be transferred from a diseased person to a healthy one by various direct or indirect routes from excreta, causing nearly 80% of sickness in the developing countries.

Mortality from diarrhoeal diseases alone contributes about 6% of the global burden of diseases. Human excreta remains one of the most serious sources of drinking water contamination. Lack of sanitation and poor hygiene are responsible for the transmission of diarrhoea, cholera, typhoid and several parasitic infections. Moreover, the incidence of the diseases linked to poor sanitation – e.g., round worm, hookworm, guinea worm, and schistosomiasis – is highest among the poor, especially amongst the school-age children. These diseases have a strong negative impact on the health and nutrition of children and their learning capacities, and contribute to significant absences from school. Trachoma can have far more devastating consequences, often leading to blindness. Women also suffer physically for want of accessible and safe toilets, as they often wait until nightfall to defecate in the open. The waiting period causes gastric disorders.

The adverse impacts of poor sanitation can extend well beyond the direct impacts on health. Health risks and epidemics from waterborne diseases can greatly reduce tourism and agricultural exports, with economic costs much greater than the cost of investments in water supply and sanitation. The wider impact of unhygienic environments drags back economic progress and erodes good governance.

The Sanitation Challenge

Whilst there is widespread recognition of the fact that improved access to safe and affordable sanitation is one of the most powerful health related interventions, scant attention has been paid to finding effective ways of promoting investments in hygiene and sanitation. The problems being associated with inadequate sanitation or lack of it, constitute a major challenge of global dimensions. There are 2.5 billion people in the world who either have no organised system of sanitation or have access only to noxious and unhygienic sanitation facilities. Globally 1.8 million people die every year from diarrhoeal diseases (including cholera) associated with contaminated water supply, sanitation and hygiene. Improved hygiene and sanitation can help reduce sickness from diarrhoeal diseases. Access to adequate sanitation facilities can reduce diarrhoea by upto 77 percent. Sanitation facilities help check transmission of many faecal/oral diseases by preventing human excreta contamination of water and soil. Epidemiological evidence suggests that sanitation is at least as effective in preventing disease as is the improved water supply.

To meet the global challenge of environmental sanitation, the whole world is speaking in one voice the language of grave concern on the difficulties that the developing countries are facing in respect of providing adequate water and sanitation services to all of their citizens. Tall promises and lofty declarations notwithstanding, the world, however, faces a crisis of frightening proportions when it comes to providing clean water and hygienic sanitation facilities to billions of people throughout the world. And, this indeed is the crux of the problematics of global sanitation challenge.

True that the recent acceleration in the pace and progress of human development centred programmes and activities has laid considerable emphasis on meeting the basic needs of the people, which include education, health, housing, water and sanitation, there, however, remain a huge number of people who do not have access to safe drinking water and sanitary toilets. No doubt, there had been some positive developments in the 1990s : about 438 million people in the developing countries gained access to safe water and about 542 million in urban areas gained access to proper sanitation. But due to rapid population growth, the number of people without access to safe water and sanitation has increased considerably. If adequate sanitation is taken to mean a toilet connected to a sewer, there is a significant lack of adequate sanitation throughout the developing world.

The state of water and sanitation services in the developing countries is pretty grim. About 2 of every 10 people in the developing world were without access to safe water in 2000; 5 of 10 lived without adequate sanitation; and 9 to 10 lived without their waste water treated in any way. There have been gains, but despite the many global commitments, notably the UN Decade for Water and Sanitation, access to water and sanitation lags far behind the milestones set in 1980s (World Bank, 2004).

The scale of the sanitation challenge in Asia is of a much higher order. In terms of sanitation coverage, Asia contains 67% of the unserved population (UNICEF, 2008). The UNICEF Global Assessment Report 2008 indicates that over 1.8 million people in Asia do not have access to adequate sanitation. In South Asia only 37% of the population has access to adequate sanitation. Some 1.4 million of the region's people still either defecate in open areas or use insanitary bucket latrines (World Bank, 2003). Over 900 million people in the region are without access to improved sanitation facilities. Many city governments are reluctant to invest in basic sanitation without addressing the broader challenges of drainage and solid waste disposal. The urban people suffer more from inadequate sanitation made worse by overcrowding in cities.

The positive role of urbanization is often overshadowed by the evident deterioration in the physical environment and quality of life in the urban areas caused by widening gap between demand and supply of essential services and infrastructure. The worst sufferers are the poor whose access to basic services like drinking water, sanitation, education and health services is shrinking. The unabated growth of urban population has made the problem of urban settlements more severe, resulting in the creation of slums and squatter settlements, causing degradation of the city environment. The task is enormous with more than two-third of South Asians lacking in access to sanitation, one-third in poverty, one-quarter hungry, one-fifth of children out of primary school and almost one-tenth of children dying before the age of five (World Bank, 2003).

Our Own Sanitation Scenario

Our own sanitation scenario is truly scaring. We still have 135 million people with no access to basic health care facilities. There are about 226 million with no access to safe drinking water. And over 650 million people lack basic sanitation facilities. Hardly 46 percent of the urban population has access to flush arrangements connected to sewerage system, 14.6 percent have water-based toilets connected to septic tanks, 13 percent have bucket latrines and the remaining 26.3 percent do not have access to any latrine facility. The coverage in rural areas is 57 percent. Nearly 70.5 percent of our population (about 650 million people) either defecate in the open or use bucket dry privies manually cleaned by scavengers.

If the sheer number of these statistics seems too impersonal, let us look at the impact of it. Every lane, by lane, park, roadside, vegetable patch, agricultural fields and railway tracks – and other open spaces are used for defecation or urination in our country by several million people. The ecological damage caused to the environment by open air defecation is our major concern in the field of environmental sanitation. Millions of people without toilet facilities infect our sources of drinking water, daily vegetables, and our surroundings. The tragic part is that most people have developed blinkers and do not acknowledge that their lives are being contaminated on a minute-by-minute basis by environmental insanitation. Providing toilet facilities can prevent this kind of environmental pollution.

Our major cause of concern is the existence of millions of bucket and dry privies and prevalence of open air defecation. It has been assessed that there are about 13 million bucket privies, of which nearly 7 million exist in urban areas. Considerable progress has been made in the country in the field of water supply but unfortunately sanitation sector has remained neglected. Even after several projects undertaken during the International Drinking Water Supply and Sanitation Decade (1981 – 91) and thereafter, about 49% of urban population in January 1997 had access to water flush toilets, of which 28% had sewerage connections and 21% had low cost sanitation facilities. By March 2000, the sanitation coverage in urban areas is reported to be 60% of which 30% are connected to sewerage and septic tanks and the balance 30% had low cost sanitation facilities. The coverage in rural areas is hardly 20%.

The use of bucket privies had led to inhuman, unhealthy and demeaning practice of manual scavenging of excreta by a particular class of people called scavengers. It results in exposing the entire locality to health hazards and degradation of environment, giving rise to a social problem. It has stigmatized the scavengers engaged in this occupation for generations. It bred the social practice of untouchability, forcing them to lead a segregated life and live in appalling conditions lacking human rights and dignity.

The non-availability of toilets, especially in rural areas, subjects women to compromise with their privacy and dignity. They go to open areas for defecation either before sunrise or after sunset which is not always safe for them. It also causes many health problems. The insanitary conditions prevailing in the country are responsible for the death of nearly half a million children each year from diseases like dysentery, hookworm or cholera. Of all the diseases diarrhoea claims the highest number of lives. Hence, every one concerned with sanitation sector has been advocating the implementation of an affordable, safe, and scavenging free system of human waste disposal which could be adopted on mass scale.

Out of our total population of 1027 million according to 2001 census, 650 million lack basic sanitation facilities resulting into high mortality and morbidity. Low sanitation coverage is primarily due to insufficient motivation/ awareness of people and lack of affordable sanitation technology. People (mostly from lower economic strata) are generally not aware of the health and environment benefits of sanitation. It is still not a felt need for them. By tradition, our society and culture value personal hygiene, but give little importance to clean and healthy community environment. Human excreta is regarded as the most hated object and anything connected with the latrine is so defiling that one is supposed to take bath immediately after coming out of the toilet and before going into the kitchen. Calling it the Faecophobia, Winblad and Simpson – Hebert (2004:100) made an interesting remark in an important publication on ecological sanitation. They wrote : "The cultural response to the fear of human excrement faecophobia has been codified into concepts of 'clean' and 'unclean' in several cultures. Hinduism is the prime example. A traditional upper caste Hindu would have nothing to do with human faeces, not even his own. This is not because they are malodorous but because they are perceived as unclean. Anyone touching them also becomes unclean. The only people supposed to handle human excreta and clean toilets are the 'untouchables'."

Response to the Challenge of Environmental Sanitation

In India, the danger signals of environmental sanitation challenge have not been clearly recognized. The realization concerning the sanitation challenge is not amply reflected in a number of writings on general environmental issues, especially those that document the damage done to environment. Most such writings aver in unanimity that the central issues concerning environmental sanitation are not getting proper government attention. This is because the nation has not realized the serious implications of a burgeoning crisis of environmental sanitation. During the last decade there has been a somewhat assuring resurgence of serious concern and consciousness. As a result of this, the Government and some environmentally conscious sanitation activists and certain civil society groups in various parts of the country are getting increasingly concerned with insanitary conditions prevailing in the country. Therefore, the need to promote environmental sanitation is being repeatedly emphasized at the highest level. However, the policy perspectives still do not dwell at length on sanitation and its linkages with environment and development. Scarce attention is still being paid to the subject of sanitation, conveniently bypassing the fact that sanitation is an integral part of environment, and by all reckoning, is an essential ingredient of development. The heart of the matter is that the serious implications of sanitation (and the absence of it) with environment have not been carefully understood.

We have not fully realized that environmental sanitation is an integral component of environmental protection, as also an essential ingredient of development. If the concern for environmental protection and development is for improving the quality of life, then a good sanitation system is one of its important indicators. Regrettably, sanitation rarely figures in different kinds of debates on both environment and development. This amounts to missing the very essence and purpose of environmental protection and development, since both aim at improving the quality of life. We have conveniently forgotten that potential health benefits from improved sanitation are huge, and investments in sanitation offer high economic, social and environmental returns (World Bank, 1992).

Despite all said and done sanitation in India still continues to be a matter of low-priority, because it is a social sector not directly connected to production sector of the country. For the fact of this perception, the significance of sanitation has not been fully appreciated. Sanitation has not been considered as a major investment priority in the country despite declaring a whole Decade as International Drinking Water Supply and Sanitation Decade and, despite the country having started the National Water Supply and Sanitation Decade Programme from April 1, 1981. And further, little thought has been given to the idea that sanitation is an important indicator of human development, which, in essence, is the central goal of development, sustainability and environmental protection. Not much credence has been given to an old observation of the World Development Report (1992) which said in clear terms that while the potential health benefits from improved sanitation services (efficient and cost-effective clean lavatories, proper sewerage, drainage and garbage disposal systems) are huge, the ill-effects of the neglect of environmental sanitation are appalling.

The fact of the matter is that despite good intentions and pious promises, the country has not been able to make much headway. The biggest barrier is the insensitivity, indifference and inaction on the part of many authorities at both the central and state levels to understand the dimensions of the problem and allocate adequate resources to solve it.

The reasons for appalling insanitary conditions in the country lie in factors like little premium being placed on sanitation; sanitation being regarded as an issue of least popular public support; inadequate attention on adopting on a wider scale the available low-cost technologies; the tie-up of sanitation to water supply (which receives more attention and greater investments); scarcity of required resources; and poor promotional measures.

The disadvantaged sections of the population which suffer most from conditions of insanitation are least organized to press for better sanitation services, perhaps because, sanitation is not as important to them as is a house, a job, electricity, water and pressing problems of bread and butter. Not many voluntary organizations have taken up the issue of sanitation. The low-prestige and recognition accorded to sanitation sector rarely enthuse the professionals to take up sanitation as a career. Amongst the professionals, the best and the brightest avoid a career in sanitation.

The neglect of sanitation prevails at the level of the policy planners too. Agencies responsible for creating a supportive environment for sanitation generally have ineffective policies. These include too much attention to water supply at the expense of sanitation. Moreover, there is often no defined approach to the problem upon which sound policy could be based.

Poor institutional frameworks with fragmented responsibilities amongst the different players in the game often contribute to inadequate sanitation promotion measures. The concerned government departments are less receptive to encouraging private initiatives of NGOs, though they can play a very powerful role in the area of sanitation promotion.

When it comes to resource allocation, both central and state governments adopt a very stingy approach; sanitation only attracts a fraction of national resources. The limited resources, however, made available are mainly channeled into hardware and not for social mobilization in respect of hygiene promotion.

The preference to hardware approach at the cost of software packages has its own share of problems. Apart from limited technology choices, the problem that is most baffling is the unified approach which prescribes universal solutions by conveniently ignoring the diversity of needs and contexts. Little regard has been paid to the fact that sanitation-related problems of different strata of society are dissimilar, and there is really no single solution; what is appropriate for urban areas may not be suitable for rural areas. To complicate the matter further the technological options offered are limited and inappropriate, and critical issues of behaviour are ignored or handled badly. Consumers are taken for a ride when technology choice is prescribed. Their socio-economic, cultural, financial and geographical circumstances are poorly understood. Too often unacceptable and affordable technologies are sold to people. In respect of low-cost technologies there is another snag : these are often viewed by many consumers as low-status technologies. Promoters often try to sell sanitation facilities on health benefits but people really want the privacy and comfort which sanitation can offer.

Poor promotion techniques fail to attract people, because promoters, in most cases, do not want to take note of what people want or believe. The concept of social marketing in the area of sanitation is yet to take its roots. There is hardly any programme which employs health educators to help understand the obstacles or to create demand. Most promoters hardly realize the fact that selling the idea of sanitation is a daunting task. Accordingly, they do a lousy job. Yet another difficulty is that promoters are often unaware of effective promotional techniques owing to lack of appropriate training and skills.

Viewed in the context of international developments in the field of environmental sanitation, India's sanitation scenario is too grim. The severity of the problem would be judged on the basis of facts regarding the existence of sanitation facilities in India's cities/towns and villages.

- Out of 5,161 cities/towns, only 232 have sewerage system, and that too not for the entire city/town. It is estimated that only 28 percent of the urban population has sewerage connections.
- In most cities and towns nearly one-fourth of the people defecate in the open and about one-fourth have manually cleaned latrine system where large number of scavengers work.
- The septic tank system is not extensive enough in Indian cities/towns inspite of its 600 year old history in the country and, barely one-tenth of the urban population can afford the system.
- According to the National Sample Survey (1973-74) almost the entire rural population barring less than one percent defecated in the open.
- Only one-third of the urban households have the facilities for the water-borne waste disposal system, while the other one-third has bucket privies and the remaining one-third houses have no latrines at all.
- It is estimated by the Planning Commission that there are over 75 lakh dry/bucket latrines in the country, of which 57 lakh are in the urban areas employing about 4,00,000 scavengers till about 1989.
- 83% to 85% of the people in the country have no latrines for their daily use.

These figures, demoralizing as they are, also refer to the magnitude of the problem of scavenging that has to be done by 400,000 scavengers.

Safe disposal of human excreta is the most challenging problem, both in rural and urban areas. The lack of modern sanitary system and the non-availability of toilet facilities with in-built system for the disposal of human excreta, health-hazard free drainage system, lack of garbage disposal facilities and the scarcity/unavailability of safe drinking water continue to be the most pressing problems. Amongst these, the problem of open air defecation in both villages and cities is the most serious and challenging one. Attached to this is the problem of bucket latrines that are to be cleaned manually by a section of people. The excreta from bucket privies often spills over during its transportation to the disposal site. Due to cleaning the privy chambers by pouring water through the squatting holes, excreta flows into drains running along the houses and, thus exposing the entire locality to health hazards and foul environment.

Human excreta is the cause of many enteric diseases, such as, cholera, dysentery, typhoid, hookwarm etc. Over 50 types of infection can be transferred from a diseased person to a healthy one by various direct or indirect routes from excreta. Therefore, safe disposal of human excreta is considered to be the primary objective of improved sanitation in order to build a healthier nation and provide a cleaner environment.

The major cause of the high rate of these diseases in India is the tradition of open-air defecation by millions of people every morning and evening, and the existence of hundreds of thousands of bucket latrines in towns and cities. It is common sight in our country to see excreta flowing through open drains due to defecation by children as well as by adults, and, also due to system of cleaning bucket privies by scavengers who discharge the residual excreta into drains which flow in front of houses.

Thus, a painful paradox persists. On the one hand a gigantic leap in the field of science and technology is ready to take us into 21st century, while on the other, the centuries old practice of manually handling and carrying night soil on the head by scavengers remains a blot on our society.

Eradication of Scavenging

Attached to the problem of environmental sanitation is the problem of the employment of those who make house-to-house clearance and collection of human excreta in buckets or baskets and carry them to disposal sites. The need for the disposal of human excreta in the Indian cities and villages has resulted in the creation of a special caste, the scavenging caste of various hues and complexions. The shame of scavenging system is that the persons belonging to the scavenging community alone are employed to perform clearing of human excreta from countless number of dry privies in both rural and urban areas throughout the country. Since the scavengers have been engaged in manual cleaning of bucket or dry privies, handling human excreta and such other avocations that the society considers as dirty, they have been looked down upon as a socially profane group, contact with whom has been considered as defiling. The stigma attached to this profession has been passed on from generation to generation and the scavengers continue to carry out this work rather unwillingly. Thus, the conservancy system is being handled by the unhappy, grudging and sullen labour. Scavengers have lost their respect for human and social values of life, which have obliged them to take up such a demeaning profession since their childhood. This appears to have affected their whole life.

Magnitude of the Problem

No reliable data is available regarding the number of scavengers involved in the occupation of scavenging, as also on the exact number of dry latrines existing in the country. However, some official estimates provide a rough idea of the situation.

A survey jointly conducted by UNDP/ World Bank on a sample of 15 lakh households in 211 small and medium towns in 22 States and Union Territories in India in the years 1983-84, furnishes some data. The findings of the survey indicated that 20.91 per cent of the households had water-flush latrines, 25.42 per cent had dry latrines and 53.67 per cent had no latrines of any type. By projecting the survey data, the dry latrines were estimated to be about 66.70 per cent in urban areas, out of which 6.0 lakh might have been converted. There was no estimate of dry latrines existing in rural areas. Moreover, the number of scavengers engaged in scavenging was also not estimated.

The Planning Commission, Government of India, therefore, constituted a Task Force in July 1989 to evolve remedial measures for tackling the problem of scavengers, which may lead to liberation and rehabilitation of scavengers through comprehensive action programmes. Besides considering many aspects of the problems of scavengers, the said Task Force also constituted a sub-committee to work out the estimated number of scavengers engaged in the servicing of dry latrines as well as the detailed State-wise estimates of the number of dry latrines in both urban and rural areas in the country.

The sub-committee of said Task Force has estimated the number of dry latrines to 76.4 lakh (54 lakh in urban areas and 22.4 lakh in rural areas) in 1989, on the basis of the growth percentage between 28th Round NSSO (October 1973 and June 1974) and 38th Round NSSO (Jan. – December, 1983). On an All-India average one scavenger is servicing about 19 to 20 households per day. The Task Force report (1991) estimated the number of scavengers for the year 1989 to be 4,00,999, out of which 3,33,779 were estimated to be in urban areas and 67,220 in rural areas. In the urban areas, 2,06,612 were estimated to be male and 1,27,167 female. In the rural areas, 53,335 were estimated to be male and 13,865 female. This rural and urban distribution of scavengers shows that 83 per cent scavengers are located in urban areas and 17 per cent in the rural areas.

The state-wise distribution shows that the problem is either non-existent, or, is numerically of a very low order in States like Assam, Haryana, Meghalaya, Manipur, Mizoram, Nagaland and Sikkim, wherefrom figures have not been reported in the estimate as above. Further, the number of persons reported to be engaged in scavenging appears to be very low in almost all the States wherefrom the figures have arrived. Since there are no specific surveys reporting the total number of scavengers on a nation-wide basis, the aforesaid figures of the Planning Commission could be swallowed with a pinch of salt.

Social Status of Scavengers : The Existential Reality

On the basis of the available material on scavenging and scavengers a very pathetic picture emerges. The social status of scavengers and the existential reality of their life and living may be described as under :

1. Scavengers constitute an underprivileged social group treated as 'Impure' and consequently ritually avoided.
2. Like 'Pariah' groups, they are treated as inferior and often regarded as less than human. They live on the fringe of society, often excluded from social contact, though their services are accepted as highly valuable.
3. They are socially ostracized because of their ritual impurity and for certain peculiar objectionable habits associated with their occupation.
4. They are backward, and often oppressed, not because they are racially, socially and culturally inferior to others, but simply because they have been assigned the lowest of the low status in social hierarchy.
5. As outcastes and rejects of the Indian society, scavengers have suffered myriad human indignities throughout the ages.

6. They have been treated like dreaded contagious disease calling for a disdainful distance and the most minimal social contact.
7. Because of their social derogation, scavengers have been compelled to live life as socially disadvantaged persons condemned to accepting their social degradation, discrimination and disadvantaged position as part of their fate.
8. Besides social backwardness, scavengers are economically backward too. Most scavenger families live in situations of poverty because of both structural and personal factors. Structural factors being the low family income on account of engagement in low paying occupations, and personal factors being their large-family size and the male members' indulgence in smoking, drinking and gambling.
9. Most scavengers are illiterate and educationally backward. In spite of the facilities for their free education, children are not sent to schools in an overwhelming number of cases, and female education is not considered to be relevant at all.
10. The economic and educational backwardness has deprived many scavenger families from taking advantage of the privileges meant for the welfare of the dispossessed.
11. Politically too, scavengers are less organized as compared with other Harijan sub-castes, the dalits and the disadvantaged.
12. While in the recent years there has been some change in the traditional occupational structure of the scavengers, a large majority of them is still given to traditional hereditary occupations.

The pitiable plight of scavenger is best summed up in the following two observations : One made by the Backward

Classes Commission in 1956 and the other by the Father of the Nation in 1927.

1. "The condition of the Bhangis* is extremely miserable and the betterment of this unfortunate section is one of the crying needs. When compared with the average income of the common people, the earnings of the Bhangis* attached to local bodies and municipalities may not be very low. Yet the condition of these people is extremely miserable. Drinking and gambling habits, lack of education and lower standards of living are responsible for many of their ills".
2. "Socially, they (Bhangis*) are lepers and economically they are worse than slaves".

National Concern on Scavenging

The national concern on scavenging was expressed in a large number of reports of the Government appointed Commissions and Committees right from the days of independence. The concern was more forcefully voiced in the Report of a Committee of the National Labour Commission on the working and service conditions of sweepers and scavengers (1968-69). The Report Said:

> "The question of working conditions of sweepers and scavengers is not merely a matter of employer/employee relationship, but an issue that has figured prominently in the social and political discussions of the country for almost a generation. The question of untouchability, caste differentiation, the stigma and repugnance attached to this occupation, the age-long suppression, privations and afflictions suffered by these communities of sweepers and scavengers involves the social conscience of the whole nation. The uplift of this class of people requires breaking of traditional attitudes and a long process of education not only of the higher castes but of sweepers and scavengers themselves, to pull them up from the

state of despair to which they have a fatalistic approach". The report further stated:

"That a section of humanity should be condemned to such degrading condition is a slur on our culture and a disgrace to our society".

Many of the national leaders have expressed their anguish over the problems of scavengers. For want of space the opinions of all of them cannot be reproduced here. However, some of the most important ones are quoted below :

"I may not be born again, but if it happens, I will like to be born in a family of scavengers, so that I may relieve them of the inhuman, unhealthy and hateful practice of carrying night soil".

– Mahatma Gandhi

"Everybody is conscious of this fact (inhuman system of carrying and handling human excreta manually), and we have discussed this many times and there is no doubt that this is a disgrace and it should be eradicated as early as possible".

– Indira Gandhi

"The continued practice of scavenging is the most undesirable form of social derogation. Under the 20-point programme (1986), we have determined to eradicate it. To do so we will require the active cooperation of all citizens and particularly of local bodies and Governments which are responsible for approving construction activity or have their own housing programmes. This will also require rehabilitation of scavengers which will mean training and resettlement in new occupations".

– Rajiv Gandhi

Reform Measures

Many social reformers have in the past voiced their serious concern for the plight of scavengers, and even worked for their liberation from human bondage. Their crusade for the liberation of scavengers did not produce any worth-while results, merely because, the very idea of the liberation of scavengers militated against well-entrenched values of the caste-ridden society. Mahatma Gandhi too had a dream for the liberation of scavengers, but his efforts produced no tangible results.

After independence, the Central and many State Governments started work for the liberation of scavengers and appointed Committees to analyse the working and living conditions of scavengers with a view to suggesting the ways of improving their lot.

These Commissions and Committees did excellent work and made extremely valuable recommendations for the consideration of the Governments, both at the Central and State-levels. Some of the major recommendations, especially of the Malkani Committee report, were accepted for 'action' by the Ministry of Home Affairs, Ministry of works and Housing, and Housing and Urban Development Corporation. International agencies like WHO, UNICEF and UNDP also chipped in to indirectly serve the cause of scavengers' rehabilitation.

While there is no denying the fact that much significant work has been done in the post-independent India in liberating scavengers from social bondage, the problem, without doubt, persists as a continuing disgrace for the nation. Many welfare schemes launched for the liberation of scavengers have not made any significant dent in the problem.

Need For Sociological Enquiry

The problems of scavenging and scavengers in the present context are not merely the problems of finding and applying inexpensive and affordable sanitation technology backed up and buttressed by ameliorative, liberative and rehabilitative policies, programmes and action plans implemented through the variety of governmental and non-governmental agencies/organizations. The sociological facets of the problem of scavenging are equally important, because they still operate with so much rigour, and characterize their stranglehold of the olden days, when caste-system was king and low-caste people were its captives. With the gradual dismantling of the steel-framework of caste apparatus, and the eventual loosening in the rigidity of the caste-borne disabilities, the depressed and dalits (scavengers included) are no doubt feeling a certain degree of freedom from centuries old social, cultural and economic bondage, but the situation of the depressed and the down-trodden has improved only marginally. The condition of scavengers is far worse, as is averred by many sociological analyses. In this context sociology of scavenging is no less important in comparison to the technology of scavenging, i.e. a new technology of sanitation with an integral component for the liberation of scavengers from the demeaning and defiling practice of cleaning bucket privies and carrying the headload of night soil.

Prof. Yogendra Singh (1992) touched the heart of the matter when he observed that "the sociological depth of this problem needs to be fully explored and action strategies evolved for rapid eradication of those social, cultural and economic factors which are responsible for the continued bondage of these people to an occupation which perpetuates their low social esteem in society".

This study, therefore, proposed to understand and examine the sociological facets of the problem of manual scavenging in the context of the plans and programmes of environmental sanitation and the liberation of scavengers.

Objectives of the Study

The study aimed at pursuing the following objectives :

1. To throw light on the genesis, growth and development of sanitation movement, including the

movement for the liberation of scavengers in the country;

2. To examine the contribution of National, International and Bilateral agencies in respect of promotional efforts in furtherance of environmental sanitation, in the liberation of scavengers, and, in the betterment of general sanitation scenario in the country;
3. To assess the impact of various schemes related to improvement in the sanitation status of country, and to assess the effectiveness of the programme pertaining to the liberation of scavengers;
4. To make an in depth sociological investigation of the problem of liberation of scavengers involving the analysis of historical, cultural, social and economic aspects connected with it;
5. To highlight the contributions of Sulabh International through the launch of a sanitation movement for the liberation of scavengers;
6. To evaluate the impact of policies and programmes evolved and implemented in the field of environmental sanitation in different States and Union Territories;
7. To present a broad perspective on the constraints of policy planning and programme execution in respect of environmental sanitation and liberation of scavengers;
8. To underline the significance of critical sociological variables in the growth and development of sanitation movement, including the movement for the liberation of scavengers;
9. To reflect upon the sociological dimensions of the problems inherent in the liberation of scavengers in the context of prevailing social order, and in

reference to technological innovations in the sanitational methodologies, and finally;

10. To present a workable programme of planned change, keeping in view the shortcomings of the past and the scarcity of resources at present.

The Genesis of Study

The idea of presenting a comprehensive sociological treatise on environmental sanitation and eradication of scavenging struck the author's mind during his intense involvement in Bhangi-Mukti programme, as also during his efforts to start a sanitation movement in Bihar. He got an opportunity to acquaint himself with the ideas and the ideology of Bhangi-Mukti and sanitation schemes launched in the early seventies. Being the Founder of a voluntary organisation (Sulabh International Social Service Organisation) engaged in the area of environmental sanitation and liberation of scavengers, the author had acquired considerable field experience and had also been in contact with large number of scavengers and the members of their family undergoing training in Sulabh Centers of training and rehabilitation, as well as with those youths who had settled in life in different trades after training. He has also been in contact with the officials of the different departments of the Central and the State Governments; with leaders of different shades of opinions and backgrounds who have had their own views on Bhangi-Mukti and sanitation movement; with various social agencies like Harijan Sewak Sangh (Safai Vidyalaya), Gandhi Smarak Nidhi, Khadi Gramodyog Sangh, Rajasthan Institute of Local Self Govt, Institute of Social Studies Trust, AFRO, Akhil Bhartiya Paryavaran & Gramin Vikas Sansthan, International Institute of Sulabh Systems, Bhangi-Mukti Sansthan, Sulabh Mahila Bal Kalyan Sansthan and other newly founded organizations engaged in this programme; with the national agencies like Gandhi Gram, NBO, HUDCO, AIIPH, NEERI, PARD, CBRI,

Roorkee, CAPART and with such international agencies as the UNICEF, WHO, UNDP & WEDEC.

Equally close has been the author's contact with the institutes engaged in research on low-cost sanitation and public health and hygiene, the technical experts working in the field, the planners and policy-makers and the action-sociologists willing to give a new dimension to the science of sociology for making it more utilitarian and relevant to the current problems of society. Through these meaningful contacts for many years, the author was acquainted with practical difficulties in executing the low-cost sanitation programme.

The author had also come to know of the working of the official agencies and voluntary organizations which had undertaken the schemes and projects relating to sanitation and Bhangi-Mukti. He had acquired detailed knowledge of the projects, and was fully equipped with field experience. These experiences, coupled with a thorough knowledge of theoretical formulations, led him to gather information about the different forms of activities in the sphere of low-cost sanitation, culminating in the liberation of scavengers.

Methodology

The author resolved that this venture should be systematically planned and meticulously executed. Consequently, he prepared a plan of action according to which information was to be collected from different States about the genesis of the sanitation movement, the various stages or phases of the movement and the present state of affairs in implementing the low-cost sanitation programme in rural as well as urban areas.

Data Collection

Contacts were established with all concerned official and non-official agencies and organizations for collecting relevant information. Official files and records were checked for gathering latest information. A number of states were covered under this contact programme.

In addition, the researches conducted by the pioneering sanitation, health and hygiene Institutes in Singur (West Bengal), Poonamalli (Tamil Naidu), Najafgarh (Delhi) and PRAI Lucknow, were surveyed and relevant information collected through various means, including personal visits to these Institutes. Demonstration units as well as working units of low-cost sanitary latrines of different designs were inspected and their efficiency and functioning was studied.

The different models of low-cost sanitary latrines developed by different agencies and research centres were also studied in order to get acquainted with their working systems and working efficiencies. The patterns of working, the area covered and the achievements made so far by agencies like the Maharashtra Gandhi Smarak Nidhi, (Poona), Safai Vidyalaya, (Ahmedabad), and Sulabh International, (Delhi), were surveyed and data obtained. The individual beneficiaries were interviewed in different parts of the country to elicit information about their experiences of the working of low-cost sanitation latrines, the difficulties they might have to face and the advantages they enjoyed. The information on defecation habits of the people in different parts of the country's villages as well as in urban habitats was collected and attempts were made to know the changes in the defecation habits of the people after the introduction of the low-cost sanitary latrines.

This aspect was analysed with a view to studying the impact of sanitation movement on the urban and rural habitats and the extent of success gained in the programmes launched by the State and Central Governments in collaboration with national and international agencies, as well as voluntary social organizations.

The scavengers and their dependents undergoing training programme, the scavengers still engaged in cleaning nightsoil and the impact of liberation programme on the social and economic life of the liberated scavengers were also relevant to the present study. This is why information about them was collected through informal interviews and through personal observations.

Use of Secondary Sources of Data

For the purpose of this study a synthetic or integrated methodology has been used by the researcher for having a deeper sociological insight into the problem under study. The methodology used in this study is based on content analysis of various official documents. It meant that the materials contained in the whole gamut of the literature relevant to the problem under study have been taken into account and their contents have been analysed to understand the extent and degree of environmental sanitation through eradication of scavenging. It is obvious that the literature dealing with the ideological and philosophical background for the genesis and growth of the movement for the liberation of scavengers from the demeaning occupation of cleaning and carrying night soil has also been taken into consideration. In addition, the written materials and documentary evidence in respect of various schemes and programmes of environmental sanitation through eradication of scavenging have also been consulted for content analysis. In this way the technique of content analysis has been of vital importance in the analysis of the problems in the present study.

In the present study the author has mainly used the techniques of observation and interview for collecting relevant information. The technique of observation was used for collecting information about the patterns and style of working of the concerned agencies and their modes of organizing their priorities in this regard. Moreover, the technique of observation has also been used to evaluate the level of achievements in the spheres of eradication of scavenging and environmental sanitation. On the other hand, the technique of interview has been used to collect relevant data from the persons serving the cause of environmental sanitation and Bhangi* Mukti in different capacities. Some of these personnel are associated with the agencies and organizations engaged in Bhangi*-Mukti and environmental sanitation programmes. The others are government officials working in different departments of Central and State Governments or in other governmental or Semi-Governmental organizations like HUDCO and Municipal Corporations. The interviews with these people had been helpful in understanding the co-ordination, co-operation and collaboration among different agencies or organs for achieving the objective of environmental sanitation and liberation of scavengers. In addition, these interviews have been of vital importance in assessing and evaluating the obstacles and hardship in the successful implementation of the programme.

REFERENCES

Government of India (1978), National seminar on Low-cost Techniques for Disposal of Human Wastes in Urban Communities, Calcutta.

—— *(1982)* Regional seminar on Low-cost sanitation, Ministry of Works and Housing, New Delhi.

—— (1982) Regional Conference on Low-cost Pour-flush Latrines, Udaipur.

—— (1961) Report of Customary Rights and Living and Working Conditions of Scavengers, New Delhi.

Issacs, Herold R. (1965), India's Untouchables, Bombay, Asia Publishing House.

Kalbermatten, J.M. (1980), Sanitation Convenience for Few and Many – Report of the International Seminar on Low Cost Techniques for Disposal of Human Wastes in Urban Communities, Calcutta

Pathak, Bindeshwar (1991), Road to freedom – A Sociological Study on The Abolition of Scavenging in India, Motilal Banarasi Das, Delhi

—— (1992), *Action Sociology and Development,* Concept Publishers, New Delhi.

—— (1992), Report of the National Seminar on Action Sociology and Liberation of Scavengers, New Delhi.

Sachchidananda (2001) People at the Bottom, A Portrait of Scavenger, Concept Publishing Co., New Delhi.

Singh, Yogendra (1992), Report of the National Seminar on Action Sociology and Liberation of Scavengers, New Delhi.

Srivastava, S.P. (1992), "Issues in Scavenging: An overview of the Problem, Programme and Policies", Report of the National seminar on Action Sociology and Liberation of Scavengers, New Delhi.

Srivastava, B.N. (1992), "State Action for the Scavengers", in Report of the National Seminar on Action Sociology and Liberation of Scavengers, New Delhi.

—— (1997), Manual Scavenging in India – A Disgrace to the Country, Concept Publishing Co., New Delhi.

United Nations (200) Report of the World Summit on Sustainable Development, Johannesburg, South Africa, UN Publications.

UN-Habitat (2003), United Nations Human Settlement Programme : The Challenge of Slums – Global Report on Human Settlements, Earth Scan Publications, London.

UNDP (2003) Achieving Millennium Development Goals in Water and Sanitation Project, Task Force Report.

Winblad, U. and Simpson-Hebert, M. (2004), Ecological Sanitation Stockholm Environment Institute, Stockholm, Sweden.

World Bank (2003), World Development Report 2003, Washington, Oxford University.

—— *(1992)* World Development Report 1992 Washington, Oxford University Press.

World Health Organisation (2003) Shaping the Future, WHO, Geneva.

—— (1987) Technology for Water and Sanitation in Developing Countries, Report of a WHO Study Group, a Technical Report Series No. 742, Geneva.

Water Supply and Sanitation Collaborative Council (WSSCC) (2002) WASH : Facts and Figures, Geneva.

---

* The word is banned under law.

Chapter - II

Scavenging Through The Ages

THE ANCIENT CONTEXT

Scavenging in India began with the rise of urban civilization. There is no evidence to prove the pattern of scavenging before the coming of the Aryan race in India. But, it is an established fact that Harappa and Mohenjodaro civilizations were developed in the pre-Aryan days. These earliest seats of civilization had urban habitats where defecation in the open was not possible. As such, scavenging was needed. But, since there is no documentary evidence on the socio-cultural life of that period and the study of the culture in that age is based on archaeological excavations, nothing can be stated definitely about the scavenging pattern. But, urbanization at that time had developed well and even bathrooms were used in those days. The presence of bathrooms and drains in the pre-Aryan days suggests that some system for the disposal of human excreta must have existed and scavenging must have persisted in some form or the other. But the absence of any record or literature stands in the way of giving any definite picture.

The Dravidian civilization was urban in character which had generally developed in the early period of slavery when a powerful tribal division of society took place and each tribe tried to live and expand at the cost of others. Cities were laid on a regular plan and the streets ran north and south taking full advantage of the prevailing winds. The houses were made of baked bricks, suggesting that there was plenty of fuelwood. The later designs in stone architecture of the Indo-Aryans also indicate the extensive availability of timber which was abundantly used for building houses. Existence of public baths, highly developed drainage and other conveniences in civil architecture indicate that surplus wealth was used more for the amenities to the people rather than building massive monuments dedicated to the glorification of the royal cult.

The sacred literature refers to the existence of slaves in the Aryan days. After the Aryans came to India and defeated the Dravidians, the latter were enslaved by the former. It is believed that these slaves were later called Shudras and assimilated into the Hindu society. Thus, the words like 'Slave', 'Chandal' and 'Dasas' came into frequent use. Those who were overpowered were termed 'Dasas' or 'Dasyus' and, later on, absorbed into the Aryans as 'Shudras'. It means that different words (or terms) were used for war captives in the beginning. It is assumed that these war captives were given dirty works like disposal of dead animals and human excreta.

The sacred literature of the Vedic and the post-Vedic periods sheds light on the system for the disposal of nightsoil. There was a particular class or section in society for doing this job. It is believed that in the beginning the work of this class or group was not hereditary and, later on, the hereditary system gradually developed. But this is certain that the inferior or weaker of them were generally war captives, who had been enslaved and were asked to do the job of removing nightsoil and doing scavenging work. One of the 15 duties for slaves enumerated in the *Naradia Samhita* was the disposal of human excreta (Nagar : 1980;9).

In the same way in the *Vajasnei Samhita* the names of the 'Chandals' and 'Paulkus' have been mentioned as the slaves who were required to clean human excreta. These references from the sacred literature also suggest that the word 'Chandal' was also used for slaves or Dasas and they were either war captives or people of the lowest stratum of society. The following observation of Kuppuswamy, 1972:204) throws some more light on the existence of scavenging in the early Hindu period in India :

"One of the earliest references to this group appears in the *Chandogya Upanishad* which used the term Chandala. The term referred to a degraded person, ranked along with the dog and the pig. The earlier Buddhist *Jatakas* (second century B.C.) also speak of the Chandals as a despised group, living outside the village and speaking a special dialect. The *Dharmashastras* declare the 'Chandals' to be the progeny of a Brahmin woman and a Shudra man. Manu also declared that these people were made to live outside the village and they were prohibited from entering the villages and towns in the day time."

The provision for the entrance of the Chandals into towns at night suggests that they were allowed to come to towns when the other people were not on the roads or in the lanes and streets and they could move without causing any problem for the highly placed people in society. So much so the Chandals or Domes (later named as scavengers) had to hang bells in their necks to keep people away from them. But the question arises as to what for they were allowed to come to towns during night hours. This might have been for cleaning toilets and for removing nightsoil without disturbing the inhabitants of the town. These statements are confirmed by the observations of Alberuni who wrote (1020 A.D) that the Dome and the Chandal as groups did the dirty work like cleaning the villages and were distinguished by their occupation.

It is evident from the above observations that Chandals were degraded people and were required to do dirty work. The same people were also called slaves or 'Dasas' and were captives or progeny of cross-breeding. But the view that all the Chandals were born of Brahmin women and Shudra men is not convincing. It is true that Chandals were following some degrading occupation and lived outside the village, but it is very difficult to imagine that persons born out of forbidden sex contacts were numerous enough to form a separate caste group, since they are to be found practically in all the villages of India. It is possible that because they were following an occupation which was despised, they were characterized by the *Dharmashastras* as equivalent to the despicable progeny of forbidden sex relations. In other words, these statements should not be taken in their literal sense in explaining the origin of these groups of people (Kuppuswamy, 1972:204-205).

It is just possible that because of their lower status, or being defeated and made captives, these people were given a specific name for identification. Those who were born out of forbidden sex relations were also not given a respectable place in society, and, as such, they were also treated at par with the slaves or captives and were called Chandals. Later on, owing to the association of the Chandals with the dirty job, they were not allowed to mix with the highly placed groups, and such restrictions, in due course, made them untouchables. Prof. Ghurye (1961) says that the ideas of purity, whether occupational or ceremonial, which are found to have been a factor in the genesis of caste, are the very soul of the idea and practice of untouchability. The same idea of purity which gave birth to the concept of pollution became responsible for declaring the Chandals and Domes as untouchables. As these people were following occupations like scavenging, doing leather work, removing dead cattle and so on, they were looked down upon as a profane group, contact with whom was defiling. So they were required to live in separate clusters outside the village (Kuppuswamy, 1972:205).

It is difficult to say whether scavengers were first regarded as untouchables because they did the dirty work, or they were asked to do dirty work because they were regarded as untouchables. But, this much is evident that in the Vedic and post-Vedic periods scavenging was done by a specific group, which was regarded as untouchable and occupied the lowest position in society. The occupation of scavenging, or cleaning nightsoil, was hereditary. It is obvious that in the beginning the task of scavenging was not undertaken willingly because nobody would have liked to remove somebody else's human excreta. It was only under compelling circumstances that a few persons (or a group of persons) had to do this job. Scavenging was a necessity to check the exposure of people to foul odour and, hence, in view of the possibility of non-willingness or refusal of people to do scavenging the dominant group made it hereditary on their part, and it became obligatory for the progeny of scavengers to remain in the hereditary occupation of scavenging. Later on, when the caste system solidified and rigidity developed in social hierarchy, the mythological sanctions were associated with the division of the Hindu society into closed hereditary status group or caste associated with a traditional or hereditary occupation.

In the later Vedic period the Shudras were treated as slaves and were required to do inferior jobs. They were treated as polluted and impure. The Shudras were no doubt recognized as a distinct order of society in later Vedic literature, but they were considered impure and not fit for taking part in religious sacrifices or reciting the sacred texts. Indeed, the *Aitareya Brahman* at one place represents the Shudra as the servant of another, to be expelled at will and to be slain at will (Tripathi, 1981:49). It also appears from this remark that marital relations or even illicit sex with Shudras was forbidden and it is just possible that in case of establishing such a relation, the progeny were regarded as polluted and called 'Chandal'.

The law books on Hinduism refer to the mixed castes which originated from such inter-caste marriages and illicit relations. They were the non-Aryans, the 'Mlecchas', 'Chandals', 'Svapakas' etc. considered even lower than the

Shudras (Tripathi, 1981:74). This establishes the view that the people born out of illicit relations were called Chandals and were regarded as inferior to Shudras. This also suggests that the word 'Mleccha' was used for being the product of socially disapproved sex relations. They were regarded impure, polluted and hence were assigned dirty jobs like sweeping and scavenging.

The names of 'Chandals' and 'Paulkus' were prevalent in the Buddha period also. The scriptures of that period also refer to the existence of the occupational groups or castes of Chandals and Paulkus. In the same way the history of the socio-cultural life during the Maurya period suggests that these castes or occupational groups were in existence. During the Maurya period Pataliputra (now Patna) was one of the five ancient towns, which had a developed municipal administration. Megasthenes has given a detailed account of the municipal administration of Pataliputra and has stated that there were six boards to discharge different functions.

But Kautilya does not mention any of these boards. He contemplates a 'Nagaraka' or 'Nagaradhyaksa' as the prefect of the town under whom were the 'Sthanikas' and 'Gopas' whose jurisdiction extended to one-fourth of the town. Consequently, the cleaning of the city must have existed. On this basis, it may be stated that scavenging existed in the Maurya period also. During this period the sanitary condition of the towns was very much improved. Chanakya, known as 'Mantri' or the principal adviser to the emperor, mentioned that every house should have a kitchen and a bathroom. He had suggested that one who defecated in the open must be fined 100 pun and for urinating 10 pun. However, those who defecated or urinated due to illness or any disability were to be forgiven.

References to Chandals are also found during the Gupta period. The Chandals were considered as polluted and social outcastes. The Chandals were regarded as social outcastes, being the only persons to do hunting and dealing in flesh. They lived away from the people and when they approached a city or market they had to strike a piece of wood, so that the other folk might avoid coming in contact with them (Tripathi, 1981:258).

It is now evident that the Chandals were treated as untouchables and they were not allowed to enter the city without announcing their entry. This provision was made to check their contacts with others. It is believed that the work of scavenging was performed by these Chandals in the cities and towns to keep them clean and free from foul smell.

THE MEDIEVAL CONTEXT

During medieval period, when the Muslims came to India, scavenging became a greater necessity. Owing to the Islamic sanction for Purdah (veil), Muslim women were secluded in the four walls of the house. The Purdah system was more strict in rich families. Men of dignity and learning considered that perfection of womanhood and chastity was attained by the observance of strict seclusion. They appear to have thought that such a seclusion was a distinction and a privilege to which female timidity and sense of pride, vanity and a false idea of chastity came to be added. This seclusion of Muslim women and their confinement in the house also gave birth to the necessity of scavenging. They could not go outside for defecation and hence the bucket privies were designed and constructed for their defecation in Purdah inside the house.

Thus, household latrines were constructed and scavenging became essential for removing human excreta. It is also important to point out that with the settlement of Muslim families in rural areas, scavenging began in these rural areas as well. It is obvious that in the past, rural population, male or female, defecated in the open space. This practice continued even in the case of converted Muslims belonging to the lower strata of society in villages. The

Muslim women from these strata went for defecation in open space, for the fact that the observance of purdah was not strictly followed by them.

But the Muslim population of foreign origin, enjoying higher social status, was very particular in observing the purdah system. Consequently, when such Muslim families settled in villages, household bucket latrines were constructed for them. The women of these families did not go outside unveiled and arrangement for defecation was made inside the house. Thus, one more dimension of scavenging was added with the settlement of these Muslim families in villages. The persons who did scavenging in towns and cities in Hindu habitats also did scavenging in Muslim households.

Moreover, those who were made captives were forced to do scavenging. This viewpoint is supported by the report submitted in 1960 by N.R. Malkani Enquiry Committee on Scavengers. Some of the families of scavengers and other Chandals and Mlecchas, who were untouchables in Hindu society, were attracted by the sanctions of equality in Islam. They embraced Islam, but even after being converted they continued scavenging. The scavengers who accepted Islam and did scavenging were called 'Halalkhors' which means 'those living on righteous, legitimate or just earnings'

This nomenclature was given to them to appreciate and praise the occupation held by them. Later on the term 'Halalkhor' was used for all scavengers irrespective of their religious identity. Akbar, being a very liberal ruler, believed in the integration of Indian society and called them 'Mehtar' which means good or better. This nomenclature was given by him as a parallel term for 'Behtar' which also means better. Thus, the term Chandala, Mleccha or Bhangi*, used till then was substituted by the term Mehtar or Halalkhor which was more high-sounding and consoling for the people engaged in cleaning night-soil and doing scavenging work.

Amrit Lal Nagar has also mentioned in his book "Nachyo Bahut Gopal" (1980) that the sub-castes of Kshatriya and Scavengers are almost the same and therefore, it can be safely concluded that some of the scavengers of today are the descendants of the Kshatriyas. Nagar refers to a book by Stanley Rice titled *Hindu Customs and Their Origins,* to prove that the so-called untouchables are usually those who lost their battles to the victors and were humiliated and the victors had forced the vanquished to perform demeaning tasks. Nagar also refers to his discussion with the well-known social worker of Lucknow, Shri Acchchey Lal Balmiki and Pandit Chintamany, of Harijan Sevak Sangh, Delhi as evidence in support of the correctness of Stanley Rice's Statement. He further wrote that Pandit Chintamani Balmiki (Raut Mehtar) gave him a book titled 'Palit Prabhakar' (meaning, Mehtar Jati Ka Itihas or A History of the Mehtar Caste). This book was written by Shri Devadutta Sharma Chaturvedi of Gazipur in 1925. The grandfather of Pandit Chintamani Balmiki, Sri Vanshiram Raut of district Gazipur had got the said book published in 1931 at his own cost. The kinds of people known by the names of 'Bhangi*', 'Mehtar', 'Halalkhor', 'Choohad' etc. have been mentioned in this book.

It is also important to point out that the caste groups existing in Indian society were further divided into sub-castes in the medieval period. The sub-groups began to be formed even within these groups and freedom of marriage and interdining was further restricted. Similarly, groups within groups appeared also among the Kshatriyas, Vaishyas, Shudras and the Chandals, and here also they affected the freedom of choice of occupation, interdining and intermarriage. This formation of sub-groups within the group developed a social hierarchy among those who were on the same level in the past. The Chandals were also divided into different sub-groups, each associated with a specific hereditary occupation and status. Thus, hierarchy emerged among the so-called untouchables or Chandals.

In the light of these facts, it appears that scavenging has been in existence in Indian society from ancient days. The rise and growth of urbanisation and the efforts for keeping the cities clean became the basis for the emergence of this occupation. It is obvious that scavenging could not be taken up by any person or group willingly and hence in view of the necessity of scavenging, the society had to impose this job upon some persons or families, who constituted a group or class of scavengers. Obviously, the society could compel only the weak and the poor to take up this job. Hence, it is certain that the weaker section of society was forced to do scavenging. Whether they were war captives, or progeny from forbidden sex relations or from some other cause, is not certain. All such views are based on only assumptions.

According to Ambedkar, these groups persisted in practicing Buddhism, and hence they were banished from the villages. This view of Ambedkar about the origin of untouchables or different groups of unclean castes is based on his own angle of vision. It is not supported by evidence. Hutton also gives his views about the 'Bhangis*' found in India. According to him not only Kshatriyas, but also those engaged in cleaning nightsoil from ancient times, i.e., 'Chandals' and 'Paulkus' and also those made captive during the war either in India or in Jerusalem or other parts of the world, are Bhangis* and Mehtars of the present age.

On the basis of the 1931 Census, Hutton gave the total number of Bhangis* in the country as 19,57,460 including 10,98,678 males and 9,18,782 females. Hence the real history of the genesis of 'Bhangis*' or scavengers cannot be given in a definite manner, but this much can be definitely stated that scavenging has been in existence since pre-Aryan days in the country.

SCAVENGING IN OTHER COUNTRIES

It is also important to note that scavenging has been in existence not only in the Indian subcontinent, but also in other parts of the world. In Europe and America, the sewerage system had been evolved after significant achievement in the fields of science and technology. Before that the problem of the disposal of nightsoil persisted and some arrangements had to be made for removing human excreta in cities and towns where open space defecation was not possible. The extract given below presents a brief account of scavenging in the western countries in the pre-sewerage days.

"Before the coming of the water-closet, the sewage of European cities was disposed of by scavengers, men who made nightly rounds of the cities (hence-nightsoil) collecting the contents of privy vaults and carting them away to nearby farming areas. Often the farmer who brought his produce to the city returned home with a load of fertilizer. This was also the custom in America, where the widespread introduction of water-closets and the sewerage system occurred only in the late 18th century. But by the mid 19th century, the scavenging system had broken down in many places. As cities grew, so did the supply of sewage. The laws of supply and demand prevailed and the bottom fell out of the night-soil maker. Farmers living near cities were saturated while for those living farther away, transportation costs were too high (Christopher, 1982)."

It is evident from the foregoing remarks that before the development of sewerage system, scavenging existed in the cities of Europe and America. With the achievements of high technology and the supply of piped water in sufficient quantity the introduction of sewerage system became possible. But it is also evident from other available accounts that in western countries scavenging was not a profession and there was no specific group or class engaged in scavenging. Consequently, unlike Indian society, scavenging was not hereditary occupation. Moreover, the social values, assigning the lowest status to the scavengers, were not found in those countries.

Christopher (1982) writes that the European farmers who came to cities for selling their agricultural commodities, loaded nightsoil during night hours and carried the same to the village for their farms. It is obvious that they did not do scavenging for earning their livelihood. They were neither employed by the city dwellers, nor received any wage for removing human excreta. They did it because they needed nightsoil to be used as fertilizer in their fields. Hence, it was at the will of the farmers to collect and remove nightsoil. It is also evident that these farmers collected human excreta at night and left the city much before sunset. Thus, it appears that the scavenging pattern in European countries was different from that found in India. In India it was the institution of caste, believed to exist by a religious sanction, which contributed to the development of scavenging as a hereditary occupation associated with a lower status.

In Muslim countries also, the system of scavenging was in operation. In Arabia scavenging existed in cities even before the advent of Islam. There is no mention in the preachings of Prophet Mohammad asking the people to construct latrines in their houses. Whereas every minor instruction or sanction declared by the Prophet is found recorded in the religious scriptures on Islam, including the traditional sayings of the Prophet, but there is no mention of the order or direction for constructing household latrine.

It is, therefore, evident that household latrines existed even before the emergence of Islam in Arabia. It is also important to point out that scavenging existed only in cities of that age, namely, Mecca, Madina, Jufa, Basra and Damascus. It may be further noted that the history of Islam, in its early days, suggests that many Arab tribes lived in fortresses. It is obvious that those living in fortresses could not defecate in open space; for them enclosed places for defecation were needed which demanded scavenging. The detailed Islamic sanctions in respect of 'Bait-ut-Khola' (latrine) are found in the sacred scriptures. These sanctions are concerned with the entry into 'Bait-ut-Khola', sitting therein and other related matters. The concept of purdah or hejab for adult girls and women in Islam also necessitated the provision of household latrines for them.

These women could not go outside for defecation and hence latrine facilities were definitely existent even in the early days of Islam. It is also significant to note that women who travelled in a caravan, also had to defecate in open space enroute to the destination at a place covered by trees or bushes. But while in town or city they used household latrines. In this way, it appears that scavenging existed in Arabia in the early days of Islam or even in pre-Islamic days. In other Muslim countries also there were household latrines for women in particular especially because of purdah system and seclusion of women.

Hence, it may be stated that scavenging existed in these Muslim countries also. But, at the same time it is significant to point out that scavenging patterns in Muslim countries were different from those in India. The concept of pollution or untouchability found in Indian society is not found in these countries. The sanction of equality and fraternity in Islam does not allow the imposition of any restriction on inter-dining, inter-marriage and social intercourse. All the believers in Islam, irrespective of their occupations, are allowed to enter the mosque and to offer their prayers together. Hence, it may be stated that while scavenging in Muslim countries did exist, the scavengers were not considered as polluted or untouchables.

It appears from the above analysis that the institution of untouchability has been a peculiarity of the Indian society. This institution or the concept of pollution associated with this occupation became instrumental in making scavenging hereditary. The development of the hereditary system, the concept of pollution and the institution of untouchability became the basic causes of exploitation of the scavengers by the well-placed and dominant castes in the Hindu society. The scavengers were not only handicapped from the economic point of view, but were also downtrodden in social, cultural and political spheres. Hence, they were exposed to atrocities, violences and victimizations. Moreover, the mythological and religious sanctions had led to the belief that it was divinely ordained to create a class of polluted and untouchable people.

This kind of a system of belief also contributed to the total submission to the customs, conventions and rituals of traditional Hindu society that discriminated people on the basis of their occupations. The untouchables did not think it desirable to protest against such a normative system, or to struggle for the improvement in their lot. Hence, not only in the Muslim period, but also during the British period they submitted to the traditional diktat of the faith-based Hindu social order. But the struggles for reformation in the nineteenth century and the development of a rational outlook under the impact of westernization aroused some degree of consciousness among untouchables, including amongst the scavengers.

The awakening of this consciousness motivated them to demand a higher status in society. When the Census was being conducted in India in 1911, the Chandals requested the Census Officer to accord them the status of Brahmin. They claimed to have a Brahmin origin and, hence made this demand. They argued that there was a similarity between them and the Brahmin caste in respect of customs, rituals and other sacramental roles. It is significant to note that this demand was made by the Chandals when their economic condition and the aggrandizement of their economic power became instrumental in their demand for an elevation in the social sphere. In reality the attainment of power in the economic sphere had raised their aspiration for achieving a respectable position in the society.

The scavengers demanded to register themselves in the Census Report as Brahmins on the basis of the claim that they and the Brahmins had the same origin. This shows that the castes engaged in the disposal of human excreta were not drawn from a single caste (Malley, 1913). By way of conclusion it can be said that scavengers of today are the descendants of the different castes of the Hindu society like Chandal, Dome, Paulkus, Kshatriya and Dhanuk etc. and came from different religious communities like Hindus, Muslims, Christians, Jews and even Sikhs.

It means that not only the Kshatriyas but also those living in the surrounding areas were made captives and were forced to carry human excreta at a distance from human habitation. Thus, these scavengers were not from one clan or religion; rather they were from different classes and religious groups. In Lucknow, Dhanuks have been found doing the job of the scavengers whereas in the other parts of this country a 'Dhanuk' is supposed to be a Kamkar, meaning that water brought by a Dhanuk is used for drinking purposes by the uppercaste of the Indian society.

THE CONTEMPORARY CONTEXT

These references to scavenging and scavengers in the ancient and medieval India notwithstanding, scavengers do not have any written history (D'Souza, 2004 : 178). In absence of an authentic history about the origin, function and features of scavengers as a sub-caste from among the untouchables, there are different interpretation and points of views about them (Chaudhary, 2001). Nonetheless, there are different speculations to explain the origin of the scavengers which are littered in the voluminous writings on India's caste system, especially in relation to the 'untouchables'. The writings of Ambedkar (1977), Crook (1986), Fuchs (1981), Ghurye (1957) Issacs (1965), Malkani (1960), Pathak (1991) and Shyamlal (1984), explain the origins of scavenging and the plight of the scavengers with vivid details of the trials and tribulations of scavengers. All of these writings have frequently used the term 'Pariah' with reference to scavengers.

Max Weber adopted 'Pariah' as a generic term for stigmatized, impure, and downgraded people. He defined pariah groups as underprivileged social groups who, through their occupations, were economically indispensable for their host societies but were treated as impure and consequently ritually avoided. These pariah groups were treated as 'untouchables', a category that includes a number of castes, collectively treated as impure and outsiders. They live on the fringe of society, often excluded from social contact with its members though their services are accepted as highly valuable and often necessary for the survival of society (Srivastava, 1997 : 15).

The pariahs in India are called by different names such as 'untouchables', 'Harijans' (a glorified term coined by Narsimha Mehta and adopted and popularized by Mahatma Gandhi) 'Exterior Castes' (as called by J.H. Hutton), "Depressed Classes", 'Outcastes' and 'Dalits'. Officially these pariah groups are included under the category of the 'Scheduled Castes'. The term 'Scheduled Castes' appeared for the first time in the Government of India Act 1935. The British Government issued the Government of India (Scheduled Castes) Order 1936, specifying certain castes, races and tribes as Scheduled Castes. The test applied for inclusion in the list was the extreme social, educational and economic backwardness arising out of the traditional practice of untouchability.

In the contemporary Indian society, the scavenging castes are known by different names in different states. In the northern India, they are Kunnas, Bhangi*, Balmiki, Chuhra, Mehtar, Mazhabi, Lal Begi and Halalkhor. In the eastern India, they are Kunnas, Hari, Hadi, Hela, Dom and Sanei. In the southern India, they are known as Mukhiyar, Thoti,

Chachati, Pakay, Relli etc. In the Western and Central India, they are known as Mehtar, Bhangis*, Halalkhor, Ghasi, Olgama, Zadmalli, Barvashia, Metaria, Jamphoda and Mela etc. But one thing is very clear that all of them stand at the bottom of the social ladder and are regarded as unclean both by Hindus and Muslims.

All these names given to the scavenger community contain an entire world of meaning – untouchable community, nature of work, station in life, and human beings despised simply because of the work that they do. And in colloquial usage, the word turns out to an insult (Mari, 1999). The most common of the plenty of terms used to describe a scavenger are 'Bhangi*' and 'Mehtar', and both are derogatory in meaning (Chamber, 1967). In Ambedkar's words, they are the lowest in a system of graded inequality (Ambedkar, 1969).

It is important to mention here that India is the only country in the world where a particular section of the society is traditionally responsible for keeping the habitation clean by removing the waste products of the society, including the human excreta. This is a disgrace to the country. Instances are not rare even today when one can see people carrying nightsoil in buckets on their heads or removing human excreta by physical touch. Who are these people? Why are they compelled to do such type of inhuman and dirty work? How long will these neglected sections of society remain under such bondage? And whether, there exists any solution to liberate them from an occupation which is an affront to their human rights and dignity? These are some of the complex questions which have engaged the writer's attention for over three and half decade (from 1970 till date).

SCAVENGER : OFFICIAL DEFINITION

Scavenging literally means cleaning up or removing filth, garbage, rubbish, dirt, etc. A 'scavenger' therefore, is a person who is engaged or employed to clean, clear, collect, remove, dispose or otherwise handle night-soil, filth, carcasses, garbage, rubbish etc The official Hindi word for the term 'scavenger' is 'Safai Karmachari' which literally speaking, means a sanitation worker. The National Scheme for the Liberation and Rehabilitation of Scavengers and their Dependents, launched in 1992, defined 'scavenger' as under :

"A scavenger is one who is partially or wholly engaged in the obnoxious and inhuman occupation of manually removing night soil and filth".

The Hindi Version of the printed text of the Scheme refers to scavengers as safai karmacharis. "Safai Karmachari" has also been defined in clause (e) of section 2 of the National Commission for Safai Karmacharis Act, 1993 as follows :

"Safai Karmacharis means a person engaged in or employed for manually carrying human excreta or any sanitation work."

It has been publicly argued by the National Commission that for the purpose of benefits under the National Scheme and for other benefits, concessions, relaxations, exemptions etc. all sanitation workers should be treated as scavengers. The National Commission pointed out that the definition of the term 'scavenger' as contained in the National Scheme is basically a functional definition and includes all those who are partially or wholly engaged in the obnoxious and inhuman occupation of manually removing night soil and filth. Further, the two words 'night-soil' and 'filth' are joined by a conjunction "and" Therefore, the functional definition of the term 'scavenger' should evidentially and obviously include all those who are engaged in manual removal of night-soil as well as filth.

Regardless of the controversy on the official definition of scavengers the term scavenger is used throughout the government records and has gained official legitimacy as an occupational category. The perspective, therefore, is that scavengers are a functional community recruited from different social groups performing this occupation (Sachchidananda, 2001). As an occupational category they are the lowest in the hierarchy of social stratification. The terms 'Safai Kamdar' or 'Safai Karmachari', or even 'Swakshakar' refer to those members of the scavenger community (now generally referred to as Balmiki Community) who are working as sanitation workers for the state.

THE MAGNITUDE OF THE PROBLEM

While there is so much noise about the shame of manual scavenging, it is ironical that nobody knows about the exact magnitude of the problem, the number of scavengers . The task force constituted by Planning Commission in July 1989, in its Report submitted on 21st March, 1991 had estimated 4,18,633 scavengers (2,60,243 males and 1,58,390 females) in the country in 1989 out of which 3,26,545 were estimated to be in urban areas and 92,088 in rural areas. In urban areas, 2,02,336 were estimated to be males and 1,24,209 females. In rural areas, 57,907 were estimated to be males and 34,181 females. In terms of the above estimates, about 78% of the scavengers lived in urban areas and only 22% were in rural areas. The estimate also reveals that majority of scavengers lived in State/Union Territories of U.P. (15.86%), Tamil Nadu (11.13%), Rajasthan (10.05%) Madhya Pradesh (8.81%), Maharashtra (8.39%), Orissa (8.03%), Delhi (8.13%), Gujarat (6.81%) and Bihar (6.67%). The State-wise estimates of the Planning Commission incidentally do not include scavengers belonging to castes other than Scheduled castes.

Now according to the Ministry of Social Justice and Empowerment, Government of India estimated number of scavengers in 2007 stands at 7,70,338, a bulk of whom belong to Uttar Pradesh, Madhya Pradesh, Maharashtra, Gujarat and Rajasthan. The table below shows the State-wise distribution.

Statewise number of Scavengers

S.No.	State/UT	No. of Scavengers Identified
1.	Andhra Pradesh	45,822
2.	Assam	40,413
3.	Bihar	12,226
4.	Gujarat	64,195
5.	Haryana	36,362
6.	Himachal Pradesh	4 ,757
7.	Jammu and Kashmir	4,150
8.	Karnataka	14,555
9.	Kerala	1,339
10.	Madhya Pradesh	81,307
11.	Maharashtra	64,785
12.	Orissa	35,049
13.	Punjab	2,988
14.	Rajasthan	57,736
15.	Tamil Nadu	35,561
16.	Uttar Pradesh	2,13,975
17.	West Bengal	23,852
18.	Delhi	17,420
19.	Nagaland	1,800
20.	Meghalaya	607
21.	Pondicherry	476
22.	Chattisgarh	3,243
23.	Jharkhand	5,750
24.	Uttarakhand	1,970
	Total	**7,70,338**

Since the statistics of the Government always remain suspect most people doubt the authenticity of these estimated figures, and some people even go to the extent of describing the official statistics as "guestimates". The people in the know of the matter felt that the population of the scavengers is about 8 lacs by conservative estimates.

THE DEGRADED SOCIAL STATUS OF SCAVENGERS

Scavengers are the most oppressed of all the Scheduled Castes. Even after about six decades of Independence, the community still remains backward economically, lowest socially in terms of caste status, powerless politically and disadvantaged occupationally. Instances of individual achievements in educational, social and economic spheres notwithstanding, the entire community is mired in penury, lack of progress and slavish submission to a variety of social evils, like widespread drinking, child marriage and mistreatment of women folk. Their social and economic conditions have not changed as desired. The occupation of scavenging and cleaning predominantly remains their responsibility. They are caught up in a psychology of the oppressed, and are having a mindset of stagnation from where they are not able to see beyond their present status. The oppressive ideology seems to have been internalized by the majority of people belonging to the scavengers' community. Most members of the community believe that their fate is divinely ordained, inscribed by God. The various cultural expressions of worthlessness used to describe them have been accepted by their community. Many members of the community still stick to their hereditary occupation of cleaning and carrying human excreta and other variety of filth, and often think this is the only station of their life. It is no wonder then that quite a good number of scavengers do not see any light at the end of the tunnel, and hence do not have the desire or determination to give up the job of manual scavenging. This is not because they like the occupation, but because there are no dependable alternatives to move on to other dignified occupations. The progressive and educated members of the community holding respectable positions experience tremendous tension and conflict in living with stigma they have been carrying by birth. As a community they are numerically weak and politically marginalized. Unlike other Schedule castes they have not been much influenced by the confrontanist ideology of self assertion through political participation to achieve social and economic equality. Being socially isolated and educationally backward, they live within shells created around them and sizeable section of scavengers sees nothing wrong with manual scavenging as an occupation that they are engaged in for generations. The profession is strictly hereditary. Women folk in scavengers' community handle human excreta and the children accompany them to work.

REFERENCES

Fuchs, Stephan (1981), At the Bottom of Society: The Harijan and Other Lower Castes, Munshi Ram Manoharlal Publication, Delhi.

Ghurye, G.S. (1957), Caste and Class in India, Asia Publishing House, Mumbai.

Hamlin, Christopher (1982), Sewerage : Waste of Resource – A Historical Perspective Scavenger 12 (3).

Hutton, J. (1961), Caste in India : its Nature, Function and Origin, Oxford University Press, Oxford.

Kuppuswamy, B. (1972), Social Change in India, Vikash, New Delhi.

Mahar, Michael J. (1973), Untouchables in Contemporary India, The University of Arizona Press, Tuscon.

Malkani, N.R. (1969), Clean People and Unclean Country, Harijan Sewak Sangh, Delhi.

Mahant, N.M. (1949), Mehtar Mimansa, Harijan Sewak Sangh, Bikaner.

Murdoch, J. (1977), Review of Caste in India, Rawat Publications, Jaipur.

Nagar, Amrit Lal (1980), Nachyo Bahut Gopal, Rajgopal & Sons, Delhi.

Pathak, Bindeshwar (1991), Road to Freedom: A Sociological Study on the Abolition of Scavenging in India, Moti Lal Banarasi Das, Delhi.

Patil, Sharad (1982), Dasa-Sundra Slavery: Studies in the Origin of Indian Slavery, Allied Publishers, New Delhi.

Sagar, S.L. (1975), Hindu Culture and Caste System in India, Uppal Publications, New Delhi.

Shyam Lal (1984), The Bhangis in Transaction, Inter- India Publications, New Delhi.

Sharma, R.S. (1958), Shudras in Ancient India Delhi: Moti Lal Banarasi Das, Delhi.

Chapter - III

SOCIAL AWAKENING : GENESIS AND GROWTH

For want of any organized reform movement in the country in the early twenties, the condition of scavengers remained deplorable. They continued to remain downtrodden both socially and economically. They were treated as the lowest among the low categories and suffered from an extreme form of economic and social backwardness. The distressing conditions of the Bhangis* did not draw the attention of social reformers, social workers or philanthropists to launch a struggle for elevating their condition from subhuman level to a human level of living. One basic reason for the absence of any initiative was the religious orthodoxy and conservatism of Indian society which did not allow any departure from the existing order of normative sanctions. The second basic reason was the domination of high caste people in social, economic and political life. This dominant section was strongly opposed to any measure which could affect its superiority and impair its control over masses. Even if there was any thinking about improving the lot of the poor it was in a generalized form and included the low caste people as well as the entire population labelled as 'untouchables'

Even the idea of generalized social reform and uplift of "untouchables" failed to assume the form of a movement in the beginning. The Indian National Congress, which is believed to be the first organized agency to initiate the idea of improving the lot of depressed classes in general, and untouchables, in particular, could not convince the rank and file in favour of launching a struggle for the uplift of untouchables.

When the history of the Indian National Congress with its controversy over social reforms is closely studied, an impression is formed that it had taken a favourable decision on the social uplift of the untouchables in its initial phase only. To be specific, the Indian National Congress in its Annual Session in Calcutta in 1917 passed resolution to that effect. The focus of the resolution was on providing social justice to the depressed classes and removing the disabilities forced by retrograde customs. An apprehension also existed about the hardships of the depressed classes. Natesan moved that resolution and Bhulabhai Desai and Rama Iyer supported it. This was all done under the Presidentship of Dr. Annie Besant (Vakil, 1985).

In the early years, the Indian National Congress was indeed not inclined to take up reform measures because it was believed that such an effort would be opposed by the privileged upper caste people and, consequently, it would jeopardise the momentum of the national struggle. The Indian National Congress decided that it should not divert its attention from its main objective of winning freedom. It feared that attention to social reforms would have a harmful effect on the national struggle. This trend continued further.

Thus, it is obvious that the achievement of freedom was the first and foremost objective of the Indian National Congress. The organization was not basically opposed to the uplift of the depressed classes. However, it could not undertake this noble cause because of the fear that the national unity might be at stake. In the beginning of the twentieth century the Indian National Congress could not adopt a clearcut policy in respect of uplifting the depressed classes. It was only in 1917, as stated above, a clear policy was adopted to this effect.

Prior to the said resolution of Indian National Congress

in 1917, some localized efforts were also made for improving the lot of the weaker section of society in general, and of untouchables in particular. One such movement was in the twelfth century in the Kannada region of South India under the leadership of Basaveswara. The movement which he launched was named Virasaiva movement. It was directed against the absolute domination of the Brahmins and the traditional normative sanctions favouring inequality on the basis of birth. Basaveswara and his associates worked for establishing an egalitarian and democratic social order and preached equality among the adherents, irrespective of their caste, creed and economic background.

The official position, which Basaveswara occupied in Mysore, proved instrumental in launching the movement and developing an organizational framework. Basaveswara was the Chief Minister and Treasurer of King Bijjala II, the Kalachuri king. This official position of the founder of the movement boosted the enthusiasm and provided moral, material and political patronage for the movement. So long as Basaveswara remained in power, the movement spread successfully. But after his differences with the ruler and his removal from this key-power position, the movement suffered a setback. The supporters of the movement worked not only for opposing the traditional social order, but also provided an organizational framework parallel to that of Brahmins to sustain and propagate their ideologies and doctrines. They preached for the worship of one God i.e. Lord Shiva and propagated Bhakti-cult for achieving the end of egalitarianism. On the basis of the available account of the genesis, growth and dimensions of the movement it appears that the movement seems to have taken birth at Kalyana. It had gathered strength and movement by about 1162 AD under the leadership of Basaveswara, Allama Prabhu, Cennabasava, Siddharama and others. During this formative phase, the organization of the movement was amorphous and nebulous. The leadership was in the hands of charismatic leaders like Basaveswara, Allama Prabhu and

Cennabasava, whose energies and resources were spent on articulating the ideology of protest and teaching the art of personal devotion and attainment of Moksha. They taught devotion to one God, Shiva – personified as Khudala Sangamma, Guhesvara, Cannamallikarjuna and Kapila Sidda Mullikarjuna in the vachanas composed and sung in praise of Lord Shiva. The element of Bhakti was emphasized by the Virasaiva saints, who asked the devotees to follow the Bhakti Marga in their quest for the Ultimate Reality (Bali, 1984:23). The egalitarian principles preached by the initiators and supporters of the movement were opposed to the doctrines of ritualistic Brahminism. The ethical codes of conduct preached and observed by Virasaivas were readily acceptable to the backward classes and downtrodden sections of society.

A vehement opposition to the movement by high caste Hindus could not be possible owing to the key power position enjoyed by the leader of the movement in the State. Consequently, there was a stir in the Brahmin-dominated social groups and radical changes started taking place. According to Bali (1984:23) the Virasaiva movement brought in its wake the transformation of institutions, changes in the legitimacy of power and changes in the core values and dominant ideologies which composed the mosaic of the social ethos of Brahmins. It prophesied the legitimacy of a new social order. The movement waged a fierce crusade against the corpus of beliefs, rites and rituals, evolved and nurtured by the Brahmins. It protested against these, as they bespoke of inequality, differential power and narrowing of the access to economic power. The cosmos of Brahminic Hinduism had come to be enmeshed in the ritual cobweb woven by the Brahmins. This alienated the Brahmins from the non-Brahmins, for the latter were denied the right and privilege to perform many of the rituals. Women suffered disabilities as they were not granted access to the performance of many rituals. The Virasaiva movement shunned elaborate ceremonialism and excessive ritualism introduced by the Brahmins. The Brahmin considered himself as the rightful representative of God on earth and regarded himself as the sole mediator or honest broker between God and man. This idea was stiffly opposed by the Virasaivas, Basaveswara and others who declared that no one had the power to act as another's spiritual representative before God. They laid emphasis on direct communication with God. They rejected the hierarchical structure of religious authority associated with Brahminic Hinduism. In their opinion, it stood between God and man and alienated them from each other. Since the most notable component of this structure is priesthood, Virsaiva Siva Saranas considered themselves as only the first among equals. Basaveswara was impatient with those for the performance of religious rites to realize god. He often exclaimed "How can devotion to Siva and adherence to rites agree?" He discarded the sacred thread. He maintained that Siva could be won over only by devotion and not either by music or by Vedic chants. He believed in full surrender of oneself to God. Many of the rituals evolved by the Brahmins were either abandoned completely or modified, simplified and refined to meet the exigencies of the time.

It is evident from the above facts that the Virsaiva movement was a revolution against Brahminism. It challenged the absolute authority of the priestly class. Ritualism was at the core of Brahmin ascendancy and the religious hierarchy was manifestation of rituals. Hence, Basaveswara and his associates struck at the very root of Brahminism and discarded all ritualistic performance. The role of a Brahmin or a priest as the mediator between man and God was rejected and personal devotion was considered the only means acceptable to God.

Chanting of Vedic slokas was not considered to be an appropriate way to be close to divine authority. Consequently, this movement attacked the traditional ritualism and the rigid hierarchical structure. The pathetic and sub-human condition of the poor and weak, in general and, of Harijans, in particular, was due to rituals and centuries-old ritualistic performances. As the Virasaiva movement rejected these normative sanctions, it provided a relief for the downtrodden. Those who accepted the teachings of Basaveswara and his associates and joined the movement could get an opportunity to improve their social standing and status through personal devotion and adherence to the philosophy of the movement. Anybody, irrespective of caste, creed or sex, could be admitted into the Virsaiva fold and escape the atrocities of the privileged section of the Hindu society.

It appears from the above discussion that the Virsaiva movement was launched to end the supremacy of the priestly class and to liberate the weaker sections from the tyrannies and atrocities of the upper class, especially Brahmins. The movement flourished in the beginning: firstly, because it provided relief to the suffering people at the lower strata and, secondly, due to the political authority and power enjoyed by the founder of the movement. But this movement was not directed towards improving the social condition of the untouchables only. It was a movement based on the ideology of devotion. He who embraced the ideology was not discriminated against by all other adherents.

But the traditional Hindu society did neither favour this movement nor recognize the change in the status of the downtrodden. Thus, the untouchables, adhering to the ideology of Virsaiva movement, enjoyed a higher status in a limited group of adherents. But they had to interact with other members of the community also and, in that case, they suffered the same humiliation and exploitation. Moreover, with the passing of time, even Virsaivas started distinguishing between touchables and untouchables and the latter were given a discriminatory treatment.

As remarked by Bali (1984:25), "Basaveswara's death, however, marked a turning point. Later, Virsaiva leaders discouraged the lower or impure castes from becoming members of the Virsaiva community. Rules for recruitment were made more stringent for those hailing from these low caste groups and seeking to become members of the movement." Thus, the partial effort for improving the lot of the impure castes suffered a setback at the hands of the supporters of Virsaiva movement at the later stage. In addition to this, the movement did not specifically aim at improving the lot of the "impure" castes; rather its programmes were partly beneficial to this section of the society. But the latent function of this movement was definitely the awakening of the consciousness among untouchables for occupying a just place in the society and for being liberated from the yokes of exploitative traditionalism.

The struggles directed towards improving the lot of untouchables as a suffering class started in the nineteenth century and gained momentum in the twentieth century. In the nineteenth century, the movement was localized, but in the twentieth century, it assumed the dimension of a national movement led by Mahatma Gandhi. The curse of untouchability was not confined to any one part of the country. It persisted throughout India, in varying intensities and rigidities. The so-called untouchables were denied even the just human rights and that too, in the name of religious sanctions and sacred commandments. Their lowest status in society was responsible for their miserable conditions in social, economic, political and cultural spheres.

P. Nirmala Bhai has rightly observed (986:54): "Harijans were strictly prohibited from having social, political, economic, religious i.e. all sorts of human rights. The pollution attributed to them was not only physical, but social, religious and psychological. Here the influence of ideology and orthodox sentiments of Hinduism was also very severe in maintaining the social life of the Harijans in India. In actual social intercourse, these sentiments of the orthodox sections of Hinduism resulted in certain distinct behavioural differences among the various castes in India. No part of the Indian society was an exception to this, especially in the light of the typical nature of Indian villages. Thus, Harijans were predominantly exposed to these behavioural differences, characteristically attributed to them on the basis of an exaggerated sense of purity and pollution by the upper castes. The sense of social isolation and restricted social intercourse had been the serious problem for the Harijans. As untouchables, they faced several other problems emerging out of the responses to, and the practices of untouchability in Indian society. These practices varied in nature and intensity of practice in different parts of India."

It is, thus, obvious that the social, economic and cultural conditions of Harijans as a community were highly miserable everywhere in the country, though differences in the degree of atrocities were also found in different regions and localities. The right thinking persons, social workers and philanthropists were moved by the miserable conditions of this community, but the extreme orthodoxy and rigidity stood in the way of their initiating any struggle for improving the lot of Harijans. The efforts at the emancipation of Harijans amounted to a declaration of war against high caste Hindus, who were dominant in social, educational, economic and political fields.

However, effort were made by social reformers from time to time to restore dignity to Harijans. These efforts and struggles appeared both at the ideological and at the action-level. These struggles proved instrumental in arousing social consciousness in favour of social justice and restoration of the just human rights to the downtrodden section of Indian society.

The pioneering effort in this respect was made by Raja Ram Mohan Roy (1777-1823). He did not launch any movement directly for the uplift of the untouchables, but his ideas of individual liberty, social justice and universalism provided a background for the ideological struggles. The Brahmo Samaj established by him revived human values, provided a platform for safeguard against exploitation and encouraged social workers and reformers to fight for the just rights of the suffering people. "The Brahmo Samaj was not a political movement : its rationalism, its universalism, its conception of the religion of humanity and its ideal of synthesis of the East and the West prepared the intellectual foundations for future national movement (Verma, 1976:36)" Thus, the organization of Brahmo Samaj helped in the promotion of a rational outlook in general, which definitely acted as an inspiring force in the arousal of ideas advocating restoration of just rights to the downtrodden section of the society.

In the same way the ideals and thoughts of Dayanand Saraswati (1824-1883), and his action-oriented efforts in the form of Arya Samaj, also occupy a meaningful place in the movement for the emancipation of the poor. It is a fact that the hereditary system of occupation and birth-oriented status in traditional Indian society formed the basis for the rigid caste hierarchy and permitted the exploitation of Harijans in the name of sacred sanction and customary obligations. Dayanand Saraswati vehemently opposed the birth-oriented occupational hierarchy and championed the cause of recognizing personal merit and individual worth. He did not challenge the sacred scriptures, but reinterpreted the religious sanctions in a rational way. He favoured Varnashrama Dharma, but gave a new explanation which served as the foundation for the development of the progressive ideas of social liberty and distributive social justice.

In this connection, Verma (1976:42) observed "Dayanand was an advocate of the Vedic Varnashrama Dharma, but bitterly criticized the iniquities associated with the caste system as practiced in India. There were tremendous evils associated with the criterion of birth as the determinant of one's caste. Hence Dayanand advocated that one's Varna was to be determined in accordance with one's psychological dispositions, qualities and actions. The idea of Dayanand was revolutionary indeed. It seemed to be a death-blow to the traditional notion of superiority conferred by birth. His own criterion of Varna, on the other hand, was democratic indeed. According to Dayanand, the theory of the four Varnas based on a psychological and occupational criterion may provide a solution of many social and functional conflicts. The criterion of Varna- "belongingness" is not birth but the psychological capacity to perform a particular function. In the social life of India Dayanand's democratic idealism was thus shown by his placing merit in place of birth."

It is thus evident that Dayanand Saraswati was strongly opposed to the birth-based system of role-differentiation and status distribution and declared that the worth or merit of an individual should serve as the basis of determining one's function in society. Instead of a rigid caste-hierarchy, Dayanand advocated for a competitive social system characterized by social mobility and free access to occupation for every individual. This idealism of Dayanand was significant in initiating the movement for the uplift of Harijans. Accepting this functional aspect of Dayanand's idealism, Mahatma Gandhi wrote, "Among the many rich legacies that Swami Dayanand left to us, his unequivocal pronouncement against untouchability is undoubtedly one."

This comment of Gandhiji suggests that Dayanand was opposed to untouchability and advocated equal opportunities and rights for all. He rejected the idea of forced occupation and declared that one should be free to choose any occupation. This idealism was, no doubt, instrumental in launching the movement for the eradication of untouchability and the movement received further strength through his organization of Arya Samaj.

In addition to the movements of Brahmo Samaj and Arya Samaj under the patronage of Raja Ram Mohan Roy and Dayanand Saraswati, the ideas and struggles of Devendra Nath Thakur (1817-1905), Keshav Chandra Sen (1838-1884), Annie Besant (1847-1933), Bhagwan Das (1869-1959), Rabindra Nath Tagore (1861-1941) and Swami Vivekanand (1863-1902) worked as dynamic forces in developing social awakening in respect of the emancipation of downtrodden people, in general, and the untouchables, in particular. Their emphasis on equality and equal opportunity and their attack on hereditary system of occupational classes and status differentiation helped break the rigidity of caste system and promote social and occupational mobility.

Bhagwan Das, for example, advocated 'equitability' as the basis for the division of labour and presented an organic social philosophy in which the four vocational classes-Brahmins, Kshatriyas, Vaisyas and Shudras would be organized in guilds (Srenis) presided over by holy men chosen from all the four classes on the basis of knowledge and experience. In this way, he discarded the hereditary system and laid emphasis on "knowledge and experience" and, thus, attached great importance to personal qualities and abilities. He also found that the western model was not suitable to Indian society.

The observations by Bhagwan Das are self-explanatory and suggest that he laid emphasis on temperamental aptitudes and, thus rejected the imposition of a vocation on the basis of heredity or birth, and caste-based division of work. It is obvious that the idealism of Bhagwan Das was favourable to the movement for the uplift of the downtrodden. In the same way Swami Vivekanand rejected the Brahmanical doctrine of Adhikarvada. On the other hand, he pleaded for spiritual equality and declared that all are equal. The exclusion of the Shudras, or the masses from the benefit of Vedic Knowledge was condemned by him.

He advocated the breaking of all the bondages opposed to the ideals of equality and democratic spiritualism. He wanted the uplift of the *Shudras* (Untouchables) or the masses, and observed: "Thus you will confer great blessing on the masses, unshackle their bondages and uplift the whole nation." It suggests that for the uplift of the whole nation, Swami Vivekanand regarded the improvement in the lot of the masses or downtrodden as inevitable. This social idealism was definitely a heavy blow to the traditional concept of Adhikaravada and the unjust rigidity in caste hierarchy of the traditional society.

The social ideas, philosophies and reformative struggles of the nineteenth century definitely brought about a radical change in the ideological world and, thus, in the social climate of the country. People started thinking in terms of an urgent need for change and for modifying and reforming the unjust and tyrannical social order. The realization that the rigid caste system was opposed to the ideals of social justice and was a negation of humanitarianism, prepared the ground for organized efforts for improving the condition of the Harijans.

As stated earlier, the political considerations and immediate objectives deferred the Indian National Congress Plan to launch a fight against untouchability. Hence, nothing concrete could be done by the Congress till 1917, when Gandhiji took its leadership. It has also been stated that association of a particular occupation with a caste or hereditary group was the basis for the development of hierarchy in society and the notion of untouchability was also a manifestation of the same establishment. Certain occupations were regarded polluted and the hereditary groups associated with them were treated as untouchable castes.

For institutionalizing untouchability, certain normative sanctions and values were also developed. Ghurye (1931) rightly points out that "the ideas of purity, whether occupational or ceremonial, which are found to have been a factor in the genesis of caste, are the very soul of the idea and practice of untouchability." Consequently, the people following hereditary occupations of scavenging, leather work or disposal of dead animals were regarded as untouchables, because they were engaged in impure and polluted occupations.

For being treated as impure and polluted, these occupational groups were known as untouchables at ceremonial level. They were prohibited from doing many things and recitation of sacred scriptures, entering temples and so on. Hence, what was needed for removing untouchability was to strike at the very system of hereditary calling. The social reformers of the nineteenth century, therefore, fought against unequal rights and advocated equal opportunities. These struggles created a stir and reformative movements were launched in different parts of the country for improving the social condition of the Harijans, so-called untouchables.

One important movement was started in Kerala by Sri Narayana Guru (1854-1928). A brief description of the movement is given by B. Kuppuswamy (1972:206) in the following words: Sri Narayana Guru established the S.N.D.P. Yogam in 1902 with his gospel, 'one caste, one religion, one God'. This organization had two important programmes; one was to uplift the Ezhavas, a depressed caste group of Kerala as part of the work of the Yogam. Narayana Guru started a number of schools and colleges throughout Kerala. Secondly, he built a number of temples and simplified rituals regarding worship, marriage and funerals. While S.N.D.P. Yogam was a secular organization with its branches throughout Kerala, Narayana Guru also established an order of Sanyasis to help the community in its religious and spiritual matters. For fifteen years, he traveled widely and helped them clean up their houses and streets and develop hygienic habits. He made them alter their diet and, thus, directed the reform to prosperity and equality.

By these two means of modern education and 'Sanskritisation', Narayana Guru sought to transform the Ezhavas from an untouchable group in Kerala to a backward caste community within a short period of 30 years. The temple entry programmes undertaken by some of the Ezhava leaders, in association with the Nayar Service Society, finally led to the temple entry proclamation in 1936. Aiyappan writes that C.P. Ramaswamy Iyer, the Dewan of the Travancore State, felt the need for consolidating the Hindus and building up the unity of the Hindus of the State as a counterpoise against the growing power and influence of the Christians. The most discontented, and at the same time the best organised among the Hindu Castes were the Iravas. Nothing was dearer to the Iravas of that generation than temple entry.

The Vykom Satyagraha was a movement started to restore social privileges to the Harijans in Kerala. It was a very important event in the State from the point of view of social reform. The Satyagraha was started in 1924 and its objective was to open the temple streets for the untouchables. The Vykom Satyagraha, though it had failed to produce any effect on the Travancore Government or the upper caste in Kerala, made untouchability a problem of national importance. It means that any immediate result of the movement was not recorded, but the struggle gave recognition to the problems of untouchables at national level. In 1936, the ruler of Travancore (now Kerala), announced the Temple Entry Proclamation, which was a significant development, keeping in view the social conditions of Harijans in those days in Kerala, in particular, and in India, in general. This announcement was unique of its type which removed the centuries-old barrier to the entry of untouchables into temples.

At the close of the second decade of the twentieth century, two dynamic personalities appeared at the national scene who fought for eradicating the evils of untouchability and improving the socio-economic conditions of the untouchables. Their struggles were not confined to a particular region; rather they were concerned with abolishing untouchability from the Hindu society. These two stalwarts were M.K. Gandhi and B. R. Ambedkar. Both of them came from western India and both were London-trained barristers. However, their family backgrounds and social strata were different.

M.K. Gandhi came from Baniya caste, whereas Dr. B.R. Ambedkar was born in an untouchable family. He was a Mahar, the largest untouchable caste in the area now called Maharashtra where this group constitutes about 10 per cent of the population (Zelliot, 1972:74). In spite of this difference in the social background and cultural heritage, both the reformers stood for equality and championed the cause of the untouchables. But their approaches were not identical. Gandhiji wanted a reformation in the Hindu society and worked for the transformation of social order, whereas Ambedkar was mainly interested in improving the lot of untouchables, within the fold of Hinduism or outside it.

According to Gandhiji untouchability had no part in this divine order - the treatment of castes below the Shudra level as unclean was not only inhuman, but harmful to Hinduism. Gandhi described it at various times as a curse, an excrescence on Hinduism, a poison, a snake, a canker, a hydra-headed monster, a great blot, a device of Satan, and a hideous untruth, (Zelliot, 1972:73). Gandhiji himself wrote in Young India (April 23, 1925) that an untouchable, "should be regarded as a Shudra because there is no warrant for belief in a fifth caste. One born as a scavenger must earn his livelihood by being a scavenger, and then do whatever else he likes. For, a scavenger is as worthy of his hire as a lawyer or your President. That, according to me, is Hinduism (Gandhi, 1937)." Thus, Gandhiji held that all Varnas are equal and there cannot be any discrimination on the basis of occupation. The work of a sweeper is as important as the work of the Prime Minister and, on the basis of occupation one cannot be regarded superior or inferior.

He favoured the law of Varna and held that a person is expected to do better in his father's occupation, because he learns it since his childhood. But this law of Varna is not obligatory nor does it compel an individual to take up the occupation of his forefathers. He should be free to take up any occupation he likes. The law of Varna prescribes that a person should, for his living, follow the lawful occupation of his forefathers, but as all occupations are equally important and equally honourable, a scavenger has the same status as a Brahmin.

The views expressed by Gandhiji about caste restrictions, caste rituals and the evil of untouchability have not remained consistent. He himself declared time and again, that whatever he uttered or wrote was based upon his 'Experiments with Truth'. This fact is found in case of his views (1920) that "Interdining, interdrinking, intermarryingÖ are not essential for the promotion of the spirit of democracy." Whereas in 1946 he wrote: "If I had my way, I would persuade all caste Hindu girls coming under my influence to select Harijan husbands." It is apparent that changes took place in the attitude of Gandhiji towards caste-based practices, with the passage of time and his experiments with realities and actual social condition. Thus, underlying Gandhi's change in attitude towards social practices was an unchanging belief in Varnashrama Dharma, the divinely ordained division of society into four groups defined according to duty – Brahmin, Kshatriya, Vaishya and Shudra. Although Gandhiji condemned the Indian caste system, he held to his belief in the traditional social order only to preserve harmony and the growth of the soul, and with it, traditional duties (Zelliot, 1972:73).

In fact, Gandhiji believed in national unity and integration of the Hindu society. He was a firm believer in nationalism and did not like the disintegration of social order. He wanted the reform of Hindu society and removal of the division of labour based on perceived superiority or inferiority. Scavenging was not an inferior job, but it was subhuman because of the fact that one hereditary group should be compelled to do scavenging. Gandhiji, therefore, opposed this practice and wanted others also to do this work. He declared that one should clean his own nightsoil and should not expect others to do it. Cleaning of nightsoil itself is not subhuman, but cleaning of nightsoil of one individual by another is definitely subhuman. Gandhiji protested against this injustice and practiced what he preached by cleaning nightsoil himself. In this way, Gandhiji wanted to bring about a change in the attitudes and outlook of the people in order to reform the caste-based unjust practices.

The approach of Dr. B.R. Ambedkar was different. His programmes were intended to integrate the untouchables into Indian society in modern, non-traditional ways and on as high a level as possible. This approach stood in marked contrast to Gandhiji's ideal scavenger "who would continue to do sanitation work even though his status would be equal to that of a Brahmin." Thus, both reformers had a vision of equality, but for Dr. Ambedkar equality meant no equal status of the Varnas, but equal social, political and economical opportunity for all. Dr. Ambedkar made his programmes to bring the untouchables from a state of dehumanization and slavery into one of equality through education and the exercise of legal and political rights. However, Dr. Ambedkar's ideology was tempered in practice by a clean perception of the tenacity of caste and tradition. He sought to awaken among the untouchables an awareness of their bad conditions and, thus, promote unity needed for effective organizations and mass action. Dr. Ambedkar advocated a separatist policy, accentuating caste distinctions in creating a society in which identities would be unimportant (Zelliott, 1972:73).

The above remarks suggest that his approach was different from Gandhiji's. He was not in favour of preserving Hindu social order and the Varna system. He wanted to raise the status of the untouchables and held the view that this improvement was possible only if the untouchables were politically significant and powerful. Hence, he wanted the oppressed to raise their educational standard so that they might know their own conditions, having aspirations to rise to the level of the highest Hindu and be in a position to use political power as a means to that end.

For raising the level of education among the untouchables to produce highly educated persons, Dr. Ambedkar founded the Peoples Education Society in 1945. The Society was the result of his programme of creating hostels for students coming from untouchable groups. He initiated this programme in the third decade of the twentieth century and his struggles in this regard led him to organize the Peoples Education Society. Thus, Dr. Ambedkar was basically interested in improving the social and economic conditions of untouchables through improvement in their educational level and political power.

Summarising Dr. Ambedkar's contribution in this field, Eleanor Zelliot (1972:78-79) remarks: "The untouchable caste has begun social and political movements before he assumed a position of leadership. His contribution was to raise these attempts to a level of effectiveness so that the caste could achieve en masse a religious conversion, build a political party and greatly increase its participation in education at all levels. Beyond his own caste, Ambedkar helped to shape the vast programme of legal rights and safeguards for untouchables which India developed. He directly influenced some other untouchable castes through the conversion movement. As a lawyer, writer and statesmen, he was an example to them all as to what an untouchable could become."

No concrete step was taken by the Indian National Congress to serve the cause of untouchables before Gandhiji assumed its leadership. Only resolutions were adopted at meetings. Since 1917, there was a change in the policy of the National Congress and it recognized untouchability as a national problem. By 1922, the Congress movement came under Gandhiji's leadership. He gave it a new life. It became broadbased. The objective of raising the social and economic status of the depressed classes was included in the Bardoli programme of the Congress. The Congress Working Committee appointed another committee to study the problems of the untouchables. Shradhanand, G.B. Deshpande, Sarojini Naidu and Yagnik were co-opted on this committee. The All India Congress Committee remained passive on the issue, doing nothing except showing lip - sympathy to them by passing resolutions.

It is also important to point out that it was not the association with the Indian National Congress that led Gandhiji to recognize the problem of the depressed classes. But his contempt for untouchability was rooted in his childhood experiences which later become an ingredient of his personality. In 1926, he himself wrote on childhood experiences about untouchability. He writes: "A scavenger named Uka, an 'untouchable' used to attend our house for cleaning latrines. Often I would ask my mother why it was wrong to touch him. If I accidentally touched Uka, I was asked to perform ablutions and though I naturally obeyed, it was not without smiling protest. I was a very dutiful and obedient child and so far as it was consistent with respect for parents, I often had tussles with them on this matter. I told my mother that she was entirely wrong in considering contact with Uka as sinful."

The childhood experience always haunted Gandhiji and when he assumed the leadership of the Congress, this experience was very much alive in his mind. No wonder, he was always keen on the welfare of untouchables and worked to ensure a just and honourable place for them in society. It is true that many social workers, reformers and ideologists projected the problem without yielding concrete results. It was only after Mahatma Gandhi took up the work of redeeming the untouchables that the work received momentum. Gandhiji called them 'Harijans' – Children of god – and organized a network of agencies to work for their cause. He also carried out campaigns against untouchability through the press, and other means. He edited a weekly journal called Harijan to highlight the magnitude of the problem and to channelize popular opinion and effort in uplifiting the untouchables. It is obvious that Gandhiji did not believe only in passing resolutions and lodging protest, but he initiated radical steps for removing untouchability.

The struggle of Gandhiji and the war against untouchability initiated by Dr. Ambedkar had some tangible effect on the conditions of the untouchables. A consciousness was gradually developing and the untouchables were also organizing themselves for a collective fight against unjust practices. The zeal provided by Dr. Ambedkar and the technique adopted by Mahatma Gandhi were manifested in the struggles of the Mahars (an untouchable community of Maharashtra). They started three movements at three different places in 1927, 1929 and 1930 in the name of Satyagraha. A depressed class conference was held in 1927 in Mahad (a small town south of Bombay).

During the Conference, a group of several thousand untouchables marched to a pond used by Brahmins. The leaders stooped and drank water and this operation was without any violence. Later, the pond was purified by the towns people. In the later part of this year, Dr. Ambedkar called another conference in Mahad and demanded the grant of right to use public water supply. At this conference the Manusmriti was burnt in public. Though this action on the part of Dr. Ambedkar and his supporters aroused a mixed reaction, this Satyagraha of Mahars at Mahad in 1927 acquired importance. The Mahars regard this Satyagraha as the beginning of their political awakening in the country.

The second Mahar Satyagraha was initiated in 1929 in an attempt to gain entry to the Parvati temple at Poona. This effort was also conducted in the Gandhian style, but it was not approved by Gandhiji or the Congress. Untouchables from several castes led by a Mahar from Poona named Shivram Janba Kamble, together with some Maratha and Brahman sympathizers joined in a four-month attempt to enter the gates of the temple complex on Parvati Hill. A song written by a Mahar for the Satyagraha related that the marchers climbed the steps to the temple gates shouting the names of Lord Shankar, Shivaji Maharaj, Cokhamela (a Mahar saint poet of the fourteenth century) and Dr. Ambedkar. Gandhi's name was not mentioned, but the technique and inspiration for the Satyagraha undoubtedly were drawn from Gandhi's teachings. Although Dr. Ambedkar's name was shouted by the marchers, he was not present.

The Satyagraha organised by Mahars in 1929 after two years clearly suggests that they were becoming conscious of their rights and were inspired by the ideas and efforts of national leaders for improving their lot. They were no longer willing to remain passive and to be exposed to injustice and exploitation. They had developed an urge to fight for their cause. But they were guided by Gandhian technique of Ahimsa and Satyagraha for achieving their goal. These struggles by Mahars created a stir in the whole of the Hindu society and its effects were not confined to Maharashtra. The Indian National Congress also expressed concern over this new trend.

An Anti-untouchability Committee was formed by Congress in March 1929. The Committee included Pandit Madan Mohan Malviya and Jamnalal Bajaj. Pandit Malviya was a prominent leader of the Hindu Mahasabha and

Jamnalal Bajaj was a Marwari businessman, who had himself built a temple for untouchables in his home town, Wardha. The committee was given the task of investigating the Satyagraha. The Committee submitted its report and did not approve of the Satyagraha. In the report, the Committee observed that about half a dozen temples were already open to untouchables and this type of Satyagraha was bound to develop bitterness obstructing future negotiations with temple trustees. The sub-committee also observed that the Satyagraha launched by the untouchable leaders of Maharashtra was not completely free from violence. This report of the sub-committee had a significant impact on the policies and programmes of the Indian National Congress in respect of such movements and struggles.

The third Satyagraha which was started in 1930 was the longest of all. It took place at the Kalaram temple in the pilgrimage centre of Nasik. The Satyagraha was organized by Ambedkar and local Mahar leaders of Nasik. Thousand of untouchables participated in the Satyagraha and tried to enter the temple and to participate in the annual temple procession. This move was opposed not only by orthodox Hindus, but also by Congressmen. Like Poona Satyagraha this effort also failed, as a result of which the untouchables under the leadership of Dr. Ambedkar developed disillusionment not only with Satyagraha, but also with the policy of the Congress.

Later Dr. Ambedkar's attitude hardened and he declared that so long as the untouchables were in the fold of Hinduism, their condition would continue to be miserable. Hence, he preferred to give up Hinduism and join some other religion. He realized that for the uplifting of the untouchables, a separate and distinct recognition was essential for them. The political significance of untouchables could not be highlighted without their recognition as a separate entity. Therefore, he demanded that untouchables should not be regarded as sections of Hindu society, but as a separate community altogether. This demand of Dr Ambedkar was accepted by the British Government, while holding the first Round Table Conference at London in 1930. Two representatives of Scheduled Castes were invited to participate in this conference and to put up their own problems. Dr. Ambedkar was one of them. This invitation amounted to the recognition of untouchables as a distinct community. Before proceeding to London, Dr. Ambedkar addressed a conference of untouchables at Nagpur in 1930 and declared that only independence of the country would bring equality to the depressed classes.

This notion of equality was not expressed in the past when struggles demanded a just place for untouchables in Hindu society. It was for the first time that the idea of equality was projected which indicated further awakening among the depressed classes. Dr. Ambedkar supported the freedom struggle of the Congress and its leaders' view that no country had the right to rule over the other. But, he said he would impress upon Congress leaders that likewise no class had the right to rule over the other class. He announced that he intended to tell Congress point blank that the proposition does not end there and that it is equally true that no class is good enough to rule over another class.

At the first Round Table Conference, the Congress could not represent itself because of Gandhiji's non-cooperation movement. At this conference, Dr. Ambedkar also demanded separate electorate for untouchables like the one given to Muslims. This demand revealed his tougher attitude towards the Congress and Hindu society. The second Round table Conference was held in 1931. It was attended by Gandhiji also. He rejected Dr. Ambedkar's demand for separate electorate and reserved seats for the untouchables. There was a confrontation between the two leaders. Gandhiji also questioned Dr. Ambedkar's bonafides and declared that he was himself a representative of untouchables.

He declared at the conference: "I say that it is not a proper claim which is registered by Dr. Ambedkar when he seeks to speak for the whole of the untouchables of India. I myself in my own person claim to represent the vast mass of the untouchables". At the third Round Table Conference Gandhiji was not present because he was in jail. The British Government announced a communal award in 1932 according to which depressed classes were given right of double vote-one in a special constituency for a modest number of reserved seats and one in the general electorate. Gandhiji reacted against this award and launched a fast unto death which created a stir in the whole of the nation. The crisis was resolved after the Poona Pact, which gave concessions to the untouchables. In the original award, they were given 78 reserved seats, while under the Poona Pact they were given 148 seats.

These political developments were creating problem for Gandhiji to work smoothly and eradicate untouchability. On the one hand the orthodox Hindus were not pleased with Gandhiji for using the word 'Harijan' for untouchables and, on the other, the untouchables were not willing to be called 'Harijan', because the term indicated patronage. Dr. Ambedkar was also opposed to this term and ultimately the term 'Scheduled Caste' was adopted which was accepted by Dr. Ambedkar as well as by orthodox Hindus. Gandhiji realized that the approach of Dr. Ambedkar was bound to disintegrate Hindu society. Hence, Gandhiji paid specific attention to the problem of untouchability and took effective steps for eradicating this evil. In 1932, the All India Untouchability League was formed under Gandhiji's inspiration. This was later called Harijan Sevak Sangh and it soon engaged itself in propaganda against untouchability and for getting open for Harijan use, roads, temples, dharmashalas, public ghats and schools. Gandhiji started a special column in the Harijan naming temples, schools, wells, etc. which were thrown open to them.

The fast unto death by Gandhiji, his organization of Harijan Sevak Sangh, and his untiring struggle to remove untouchability bore fruit in future. A number of events and developments took place which were definite indices of growing social awakening in the country. The high caste Hindus were also realizing the need for a change in attitude, outlook and caste-based practices. Even in South India, where untouchability was practiced in the most rigid form, the signs of change started appearing. In 1932, an effort was made for the entry of untouchables in the famous Guru Vayur temple at Malabar in Kerala. The movement was organized and led by Sri Kelappan Nair and he was supported by a number of right thinking people and national leaders. The struggle was initiated with a moral strength assisted by rational justification and consequently the ruler of the territory, along with other social reformers, declared the temple open to Harijans on November 1, 1936. This declaration was a major step in the direction of eradicating the evil of untouchability and in recognizing the just and human rights of the depressed classes in India. The above facts indicate that the period between 1917 and 1947 was one of transition.

Meanwhile, the very social order of the traditional Indian society was gradually changing under the impact of Western culture, rational ideologies, nationalist movement and political developments. The views of Indian leaders were no more unheard of and unattended to by British rulers. The echo of the voice of national leaders was also heard in the British Parliament. The British rulers had started discussions with Indian leaders on issues of national importance. Dr. Ambedkar, when invited to the first Round Table Conference was under the impression that Swaraj was to be granted to the Indians. Hence, his tone at the conference was a bit harsh in criticizing Hindu leadership and in expressing doubts about the sincerity of Congress leaders about the problems of untouchables. In the beginning, he not only favoured the said cause but also organized movements for the entry into temples.

However, later on, he abandoned this idea, He thought that the basic problem of the untouchables was not social, but political. Hence, he became more concerned with the recognition of untouchables at the political level as a separate and distinct entity. Consequently, he demanded separate electorate. To him, the only concern was the interest of the untouchables. He was not bothered about the whole of the nation. On the other hand, Gandhiji and the Indian National Congress were concerned with the problems of the entire nation, achievement of independence was the most important among them. The problems of nationalism, social justice and reconstruction of social order were no less important for Gandhiji. He did not favour one strategy at the cost of the other. He was also opposed to untouchability and sought to uplift the depressed classes. He realized that untouchability was deeply rooted in Hindu society and had persisted for centuries. Hence, this evil could not be eradicated at one stroke.

In other words, Gandhiji favoured the idea of achieving the goal, step by step, and not by revolution. He realized that in case an attempt was made to change the social order at once, there would be a chaos: for the men in power were yet to change their mind. Temple entry, for instance, was a very delicate issue. Dr. Ambedkar and his associates wanted all the temples open to untouchables immediately. Gandhiji had realistic approach. In *Gandhi Shikshan,* he wrote that it was not possible that the Antyajas should have the right to enter all the existing temples. As long as the law of caste and Ashram had the chief place in Hindu religion, to say that every Hindu could enter every temple was a thing that was not yet possible. In fact, both Gandhiji and Dr. Ambedkar were eager to abolish the evil of untouchability, but their approaches were different.

Dr. Ambedkar was a radical because he had a limited interest in uplifting untouchables. He at times threatened to give up Hinduism and sometimes he demanded a place of equality for untouchables in caste-oriented Hindu society. At one stage he opposed the idea of separate electorate and at another he demanded it. Gandhiji was a realist and he had to take many interacting forces into consideration. He wanted to save Hindu society from disintegration, through peaceful changes in the social order. On many occasions, Gandhiji was bitterly criticized by Dr. Ambedkar, but he did not submit to the pressure tactics. In the process, Dr. Ambedkar had to encounter confrontation with a top-ranking leader like Mahatma Gandhi. The encounter symbolized certain trends in India. Gandhiji was aware that the political decisions were bound to be largely influenced by the political radicals; though not to be a social reformist was a key to successful leadership. This was evidenced by the role Lokmanya Tilak, the immediate predecessor of Gandhiji played like a skilful leader and had to make compromises with so many forces. Consequently, his policies and approach to untouchability never became revolutionary. Without taking into account the representations of the political leaders, Gandhiji did not take any decision. The Congress and Gandhiji had several limitations caused by number of factors. Perhaps, Gandhiji appeared to be an orthodox and egoistic Hindu and did not like his spokesmanship shared by any other leader. By becoming a real Hindu, he saved the Indian subcontinent from falling into the hands of communal Hindus. If the social reformist in Gandhiji would have dominated his counterpart, the said achievement would not have been realized.

The radicals, including Dr. Ambedkar, did not realize the relevance of the Gandhian approach to social and political problems of the country. On the other hand, they disdained Gandhiji's traditional outlook and modes of behaviour. They considered his philosophy outdated and rejected his programme which was based primarily on a concern for the rural masses. In any case, they had no interest in a drab reform programme which could neither stimulate their intellect nor excite them to revolutionary action.

The foregoing discussion does not seem to place the situation in proper perspective. Confusions about Gandhiji's approach arose out of the complex situation due to conflicting forces and ideologies in the country. The difficulty with Gandhiji was that he was not a sectarian, communal or caste leader; rather he was the leader of a nation which was very complex in nature and structure. The interest of the nation as a whole could be preserved when any one section was not encouraged to wage a war against the other. This task was gigantic which only a charismatic personality like Gandhiji could do without damaging the structure. He did not support the revolutionary and reactionary ideas and programmes of Dr. Ambedkar. But this much is certain that these two currents of thoughts and action plans of Gandhiji and Dr. Ambedkar influenced each other. If, on one hand, Dr. Ambedkar's strategy drew the attention of Congress leaders to the more specific problems of the untouchables, the approach of Gandhiji and his struggles for the uplift of the untouchables kept most depressed classes with the Congress.

The currents and cross-currents at the national scene contributed to the creation of social awakening, the genesis of which can be traced to the nineteenth century. But, it grew and developed in the twentieth century under the leadership of Gandhiji, who mobilized the Congress, and as a consequence of which other national leaders also realized the need of removing untouchability and granting equality to all in the country. The very Preamble of the Constitution of India declares no distinction on the basis of caste, creed, race or religion. The untouchability Act of 1955 made untouchability a cognizable offence throughout the country. In this way, the movement started by Mahatma Gandhi bore fruits after independence.

The foregoing discussion presents a brief account of the genesis and growth of social awakening and the uplift of the depressed classes, or untouchables, as a whole. It is true that scavengers are included in the category of the untouchables and the efforts and struggles made for removing untouchability and for improving the lot of untouchables were good for the scavengers also. But, these efforts were not concerned specifically with the Bhangis*, who were regarded as the lowest even among the untouchables. The occupation with which they were traditionally associated was very much humiliating and sub-human. The customary practice compelled them to clean nightsoil of others and to carry head-loads of human excreta for disposal. On account of being engaged in this demeaning occupation, they were regarded as polluted even by untouchables.

Various movements launched in different parts of the country were concerned with different groups of untouchables. But, none of them was concerned with the 'Bhangis*'. It was Gandhiji, who first paid attention to the pathetic condition of the 'Bhangis*'. As stated above, the seed of innovative ideas against untouchability was sown in the heart and mind of Gandhiji while he was still a child.

Scavenging and the 'Bhangi*' figured prominently in many of Gandhi's pronouncements and actions. He himself cleaned the latrine at the Calcutta Congress of 1901 and has recorded it in his Autobiography. He often used the metaphor of the mother's cleansing work for her child as a counterpart to the Bhangi's* work for the society. Sanitation work at the Gandhi Ashram was done by all members as a means for demonstrating the honourable nature of these essential duties. In later years, Gandhiji stayed sometimes in the colonies of the scavengers. The Late Prime Minister Mrs. Indira Gandhi, while participating in a debate in Parliament, disclosed in 1983 that she herself had to clean the human excreta in turn of the inmates of the Gandhi Ashram while she was living there. These facts clearly reveal the inner feelings of Gandhiji for the scavengers and his inner urge to do something for the sufferings of this occupational group. His love for the scavengers and his ardent wish to raise their social status are a reflection from his following declaration in the last days of his life, "I would rejoice to think that we had a sweeper girl of a stout heart, incorruptible and of crystal purity to be our first President....assisted in the discharge of her duties by a person like Pandit Nehru."

Gandhiji was opposed to any distinction or discrimination on the basis of occupation or profession. To him all work was equally important. For expressing equality in this context, he frequently mentioned scavengers and their occupation. On one occasion he wrote "a scavenger has the same status as a Brahmin." Ten years later he again asserted, "....a scavenger is as worthy of his hire as a lawyer or your President." Gandhiji was so anxious to change all traditional notions and assumptions regarding scavenging that he made cleaning and scavenging a precondition for joining his Ashram.

S. Chandrashekhar describes this fact while writing about his meeting with Mahatma Gandhi before his visit to Columbia University in U.S.A. in 1941 for his doctoral work. It was suggested by his father to meet Gandhiji before his departure for the U.S.A. Describing his meeting with Gandhiji, he wrote that Gandhiji talked about the question of public sanitation and hygiene. The Indian tradition had described the work of sweepers and scavengers as unclean, and the Harijans had to do it. The first test of any one desiring to become a part of the Ashram and a follower of Gandhi was to do some scavenging work. The difficulties of the scavengers engaged his attention to such an extent that he would write about their problems in almost every issue of "The Harijan". He said, "I may not be born again but if it happens I will like to be born in a family of scavengers so that I may relieve them of inhuman, unhealthy and hateful practice of carrying head-loads of human excreta."

The above facts reveal the concern of Gandhiji with the problems of untouchability and poor condition of the 'Bhangis*'. He not only propagated the idea of 'Bhangi*-Mukti' but through his plan of action and his own deeds made pioneering and valuable contribution to this sphere. He regarded it unjust that only the 'Bhangis*' should be compelled to do scavenging. There should be no hereditary obligation for scavenging and members of higher castes should also do the sanitation work. The initiative taken by Gandhiji proved instrumental in creating awareness for the liberation of scavengers.

In addition, the persistent efforts and struggles by Gandhiji made the scavengers also conscious of their own exploitation in traditional social system and they were also mobilized to think in terms of changing their status. Thus, along with the improvement in the social image of scavengers, the self-image also reflected signs of improvement. The task undertaken by Gandhiji influenced many of his associates and, thus, a team of dedicated social workers was prepared which took up the task of sanitation and improvement in the lot of scavengers. The efforts made for the liberation of scavengers also created consciousness about the sanitation programme. It is also important to point out that on the one hand Gandhiji fought for the rights of the untouchables at national level and in political perspective, and, on the other, he remained engaged in actual social service arousing social awakening in this regard.

As described earlier, Gandhiji took up the cause of the scavenger in 1901, when he cleaned a dirty latrine on the occasion of Calcutta Congress. It was a symbolic step, for it denounced this traditional notion by his own deed that scavenging was an exclusive job of the scavengers. He cleaned the latrines and even after that he was not declared an untouchable by orthodox Hindus. If scavenging made a person polluted, he should also have been thus treated. This act of Gandhiji gave a heavy blow to the institution of untouchability and ideology of pollution. Gandhiji went a step further when he established his famous Sabarmati Ashram in 1918 and people started joining.

He made cleaning and scavenging as a pre-condition for entry into his Ashram. The people who did scavenging in the Ashram did not become untouchables in society and, thus, Gandhiji set an ideal of doing the job of scavenging without being polluted or being untouchable. With his words and deeds, he created a climate favourable to the cause of the scavengers and aroused a spirit among many of his associates to work for uplifting the suffering of this class. In fact, the ideas and work of Gandhiji covered a wide range of issues - social, economic, political and cultural. His associates and disciples were influenced at different levels and in varying dimensions from time to time.

Every dedicated worker and follower of Gandhiji chose one dimension and worked in a limited perspective with sincerity and devotion. 'Bhangi*-mukti' was also an important dimension of Gandhian philosophy and action-oriented strategy. Some of his followers and associates took up the social work and devoted their energy and time to serve the noble case of the 'Bhangis*' in different parts of the country. As Gandhiji had established his Sabarmati Ashram in Gujarat, the centre of 'Bhangi*-mukti' remained more concerned with Gujarat and Maharashtra States.

Thus, the sanitation movement and struggle for uplifting 'Bhangis*' were taken up by these dedicated workers in different phases. One of his disciples and followers, Vithal Phalke, took up the task of uplifting scavengers in his own Ashram in 1921. This Ashram was established by Vittal Phalke at Godhra in Panch Mahal district of Gujarat. That the Ashram was popularly known as 'Bhangi* Basti' suggests that the Ashram was set up with a view of uplifting the scavengers and serving their cause. The 'Bhangi* Basti' laid emphasis on health and sanitation along with denouncing this belief that cleaning or scavenging was to be done only by the scavengers. In this way the struggle became bi-dimensional, in the sense that it aimed at the liberation of scavengers and it also took the form of sanitation movement.

One of the leading disciples of Gandhiji in the cause of Bhangi* Mukti was Appa Saheb Patwardhan. He was very much impressed by Gandhiji's approach and joined Sabarmati Ashram in 1919. He evinced keen interest in Bhangi* Mukti programme and struggled throughout his life for this cause. During his participation in Khadi movement in Maharashtra, Appa Saheb Patwardhan observed that the sanitary conditions were very poor in the state. He was also moved by the pathetic condition of the scavengers, who came from other States to do scavenging work. These observations and experiences, combined with Gandhian ideologies and programmes for the uplift of the scavengers produced an abiding impact on Appa Saheb and he took up the cause of the sanitation work as the objective of his social service programmes. For liberating the scavengers from the demeaning work of cleaning night soil, he designed 'trench latrine' and propagated its use in the surrounding areas. He also developed the idea of using mobile 'trench latrines' in the agricultural fields.

Consequently, the mobile trench latrine design introduced by Appa Saheb was significantly helpful in the Bhangi*-Mukti programme. While he was in Ratnagiri jail in 1932, he observed that the work of scavenging was given by jail authorities to the prisoners of the lower class of untouchables. Other persons confined in jail were not asked to do scavenging work. Thus, the prison authorities also observed the discrimination on the basis of traditional values. Appa Saheb did not like this practice and decided to do scavenging work in jail. He requested the jail authorities to allow him and his associates to share the work of the scavengers and to clean latrines in jail. But the authorities did not permit it. Appa Saheb demanded a change or amendment in the Jail Manual and gave this notice to the authorities that he would go on fast unto death if the necessary amendments were not made. The authorities refused to concede this demand. Hence, Appa Saheb went on fast unto death in Ratnagiri jail. When Gandhiji came to know about these developments, he immediately took up the matter with the Viceroy of India and demanded immediate acceptance of the demand of Appa Saheb. Gandhiji also made it clear that in case Government did not take up this matter in the right perspective, he would also join Appa Saheb and would launch a fast unto death. The seriousness of the issue and the gravity of the situation was sensed by the Government and necessary amendments were made in the Jail Manual to concede the demands of Appa Saheb. Ultimately, Appa Saheb and his companions were given the task of cleaning latrines.

In the following years Appa Saheb did exemplary work in the field of sanitation and Bhangi*-Mukti programme. He designed various types of latrines suited to different situations and introduced them to common people. The basic objective of Appa Saheb was to liberate the scavengers and to check environmental pollution through improving sanitation. These programmes undertaken by Appa Saheb and his associates as well as their organizational activities under the dynamic leadership of Gandhiji got a welcome boost. A stir was created in the society and along with the arousal of awareness in respect of sanitation, this realization was also gaining momentum that scavengers should be liberated from this demeaning job.

It is evident that the struggle for the genesis and growth of social awakening has a long history. The unorganized efforts and ideas appearing at different places and at different stages of time gradually got crystallized and well organized and took the form of a well-organised movement. This movement spread far and wide in the country, though certain areas were more intensely covered in the beginning.

But the network of Indian nationalism and the work of Indian National Congress under the leadership of Gandhiji made this movement nationwide. Whatever happened in one area had its repercussions and impacts on the other. In this way a social awakening in respect of sanitation and scavenging originated, grew, developed and generated a zeal for the completion of the said goal.

REFERENCES:

Ambedkar, B.R. (1943), Gandhi and the Emancipation of the Untouchables, Thacker, Mumbai.

Ambedkar, B.R. (1953), What Congress and Gandhi have done to the Untouchables, Thacker, Mumbai.

Agarwal, L.B. (1934), The Harijans in Rebellion, Taraporewala & Sons, Mumbai.

Beteille, Andre (1969), Social Inequality, Penguin Books.

Bose, N. K. (1975), The Structure of Hindu Society, Orient Longman, New Delhi.

Bali, A. P. (1984), "Organisation of the Virasaiva Movement", in M.S.A. Rao (ed.) Social Movements in India, Manohar Publications, Delhi.

Chintamani, C. Y. (ed.), Indian Social Reform, Thomson & Co., Chennai.

Das, J. R. J. (1968), Bhangi Kasht Mukti and Bhangi-Mukti, Harijan Sewak Sangh, Delhi.

Gandhi, M. K. (1954), The Removal of Untouchability compiled by B. Kumarappa, Nav-jeevan Publishing House, Mumbai.

Kuber, W. N. (1973), Dr. Ambedkar : A Critical Study, People's Publishing House, New Delhi.

Kuppuswamy, B. (1972), Social Change in India, Vikas, Delhi.

Lynch, O. M. (1969), The Politics of Untouchability, Columbia University Press.

Mahar, Michael J. (1973), Untouchables in Contemporary India, The University of Arizona Press, Tuscon.

Malik, S. C. (1977), Dissent, Protest and Reform in Indian Civilization, Indian Institute of Advanced Studies, Shimla.

Natrajan, S. (1959), A Century of Social Reform in India, Asia Publishing House, Mumbai.

Oommen, T.K. (1990), Protest and Change, Sage Publications, New Delhi.

Patwardhan, S. (1973), Change Among India's Harijans, Orient Longman, Delhi.

Ram, Jagjeevan (1980), Caste Challenge in India, Vision Books, New Delhi.

Vakil, A. R. (1985), Reservation Policy and Scheduled Castes in India, Ashish Publication House, New Delhi.

Verma, V. P. (1967), Modern Indian Political Thought, Allied Publishers, Mumbai.

Zelliot, Eleanor (1972), "Gandhi and Ambedkar : A Study of Leadership" in J.M. Mahar (ed.) The Untouchables of Contemporary India, University of Arizona Press, Tuscon.

---

* The word is banned under law.

Chapter-IV

MOVEMENT FOR THE EMANCIPATION OF SCAVENGERS

The discussion contained in the preceding chapter shows that while several social reform movements were launched against the practice of untouchability and for the improvement in the economic/social/ritual status of the untouchables, no social movement talked specifically about improving the working and living conditions of the people of the scavenging caste. Consequently, they continued to remain down-trodden, both socially and economically. They were treated as the lowest among the low-categories and suffered from extreme form of economic and social backwardness.

The distressing conditions of the Bhangis* did not draw the attention of social reformers, social workers or philanthropists to launch a struggle for elevating their conditions from subhuman level to human level of living. The basic reason for the absence of any initiative was the religious orthodoxy and conservatism of Indian society which did not allow any departure from the existing order of normative sanctions. The second basic reason was the domination of high caste people in the social, economic and political life. This dominant section was dead opposed to any measure of amelioration in respect of scavengers.

It is true that scavengers are included in the category of the untouchables and the efforts and struggles made for removing untouchability and for improving the lot of untouchables were applicable for scavengers also. But these efforts were not specifically concerned with them, though everybody sympathized with their pitiable working and living conditions. It was Gandhiji who paid attention to the pathetic condition of scavengers.

Scavenging and scavengers figured prominently in many of the pronouncements and actions of Gandhiji. To him the scavenger was the symbol of irony and humiliation. To him it sounded ironic that one on whom society depends for its cleanliness, sanitation and health is regarded as an untouchable and relegated to a subhuman life. He had argued that the scavenger was not only a victim of humiliation, but also a reminder of the humiliation that society had brought upon itself by dehumanizing those who performed one of the most essential services. He, therefore, identified himself with the scavenger. He challenged the concept of untouchability and pointed out that the services of a scavenger were comparable only to the services of the mother. He wanted all discrimination against the scavengers to end. He desired the scavengers to receive equality of status and opportunity. He also wished the scavengers to realize that scavenging was a fine art; and, therefore, scavenger should know the scientific way of constructing latrines, cleaning, disinfecting, and applying of the best methods of disposal of nightsoil. To him, sanitation was a science and an art. The importance that he attached to the work of the scavenger is evident from the fact that he wanted everyone, who joined his Ashram to begin by participating in the work of scavenging as part of his sacrificial bread labour.

Gandhi was so deeply moved by the pitiable position of the scavengers that he once remarked, "I may not be born again, but if it happens, I will like to be born in a family of scavengers so that I may relieve them of the inhuman, unhealthy and hateful practice of carrying head loads of nightsoil". His concern for scavengers can be easily imagined from his view that if there was rebirth in store for him, he wished to be born a Pariah in the midst of Pariahs, because there he would be able to render more effective service to them and also be in a better position to plead with other communities on their behalf.

Scavengers for Gandhiji were a crucial part of the society. Without the scavengers, he said, the society would cease to exist. He went to the extent of claiming that it was the Bhangi* who enabled society to live. The importance of scavengers for society can easily be gauged from the fact that Gandhi drew a comparison of the role of a scavenger in society with that of a mother. A scavenger does for the society what a mother does for her baby. A mother washes her baby of dirt and ensures his health. Likewise, the scavenger protects and safeguards the health of the entire community by maintaining sanitation for it. In Gandhiji's view, the scavenger should be accorded a higher position than that of a Brahmin in the society, because he thought that the service of the scavenger is more useful and important than that of a Brahmin. But without the former the latter could not be. The Brahmin's duty is to look after the sanitation of the soul, the scavenger's duty is to look after the body of the society. The scavenger constituted the foundation of all services, opined Gandhiji. The same view is held by Gandhiji when he compared the services of the scavenger with that of a doctor. The doctor's profession helps only patients but the scavenger performs more useful and sacred duties than the former. If the doctor stops practicing, the sick will perish, but should the conservancy services stop, the world will suffer.

Though the scavengers are essential for the existence of society, Gandhiji was pained to witness their wretched condition in society and he exclaimed that the "Indian society has branded the scavenger as a social pariah, set him down at the bottom of the scale, held him fit only to receive kicks and abuses, a creature who must subsist on the leavings of the caste-people and dwell on the dung heap". Socially scavengers are like lepers, economically they are worse than slaves, and religiously they are denied entrance to places we call houses of God. They are denied the use of public roads, public wells, public taps, public parks and the like, and in some cases their approach within very sight is an offence. They are relegated for their residence to the worst quarters of cities or villages, where they practically get no social services. Caste Hindu lawyers and doctors will not serve them as they do the other members of society. Brahmins will not officiate at their religious functions. They are too downtrodden to rise in revolt against their oppressors. They do not today possess even good brooms or other suitable means for the removal of nightsoil. The sight round the scavengers' quarters is no better than a cesspool of dirt. This picture was probably presented by Gandhi with a view to drawing our attention to the wrong done to scavengers by the society. Only by realizing our faults can we rectify it, Gandhiji emphasised.

Gandhiji wanted complete removal of the practice of untouchability, and he advised the Indians to change their attitude towards scavengers. Such a changed attitude for scavengers would be nothing but a better understanding of Hinduism, and it would do a great service to Hinduism since untouchability was compelling many of them to abandon Hinduism and embrace a religion in which there were no caste barriers. He said : "I want to remove untouchability, because its removal is essential for Swaraj and I want Swaraj. But I would not exploit you for gaining any political ends of mine. The issue with me is bigger even than Swaraj. I am anxious to see an end put to untouchability because for me it is an expiation and a penance. It is not the untouchables whose Shuddhi would be absurd, but my own, and that of the Hindu religion. Hinduism has committed a great sin in giving sanction to this evil and I am anxious if such a thing as vicarious penance is possible to purify it of that sin by expiating for it in my own person". He wanted to uplift Hinduism, of which the so-called untouchables are an integral part. As he put it, "I am pained when I see a single scavenger driven out of the fold of Hinduism. But I do not believe that all class distinctions can be obliterated. I believe in the doctrine of equality as taught by Lord Krishna in the Gita. The Gita teaches us that members of all the four castes should be treated on an equal basis. It does not prescribe the same Dharma for the Brahmin as for the scavenger. But it insists that the latter shall be entitled to the same measure of consideration and esteem as the former with all his superior learning. It is, therefore, our duty to see that the untouchables do not feel that they are despised or looked down upon. Let them not be offered leavings from our plates for their subsistence. How can I accord differential treatment to any person, be he Brahmin or scavenger, who worships the same God, and keeps his body and soul pure and clean? I for one would regard myself as having sinned, if I gave to a scavenger unclean food from the leavings from the kitchen or failed to render him personal assistance when he was in need. Let me make my position absolutely clear. While I do hold that the institution of untouchability as it stands today has no sanction in Hinduism, Hinduism does recognize untouchability in a limited sense and under certain circumstances. For instance, every time that my mother handled unclean things she became untouchable for the time being and had to cleanse herself by bathing. As a Vishnava, I refuse to believe that anyone can be regarded untouchable by reason of his or her birth, and such untouchability as is recognized by religion is by its very nature transitory, easily removable and referable to the deed, not the doer. Not only that; just as we revere our mother for the sanitary service that she renders us when we are infants, and the greater her service the greater is our reverence for her, similarly the scavengers are entitled to our highest reverence for the sanitary service they perform for society". Gandhiji was of the firm view that not till the distinction between the Brahmin and the scavenger was removed, would our society enjoy health, prosperity and peace, and be happy.

To remove the stigma of untouchability from the scavengers, Gandhiji advised that everybody should do his own scavenging, hoping that scavenging then would become a duty and not remain a profession. He made it a rule that any person desiring to join the Sevagram Ashram, would, in the beginning, have to clean his own toilet. He held the view that everyone must be his own scavenger. Evacuation is as necessary as eating, and the best thing would be for everyone to dispose of his own waste. If this is impossible each family should do to its own scavenging. "I have felt for years that there must be something radically wrong, where scavenging has been made the concern of a separate class in society. We have no historical record of the man, who first assigned the lowest status to this essential sanitary service. Whoever he was, he by no means did us any good. We should, from our very childhood, have the idea impressed upon our minds that we are all scavengers, and the easiest way of doing so is for everyone who has realized this, to commence bread-labour as scavenger. Scavenging, thus intelligently taken up, will help one to a true appreciation of the equality of man". And again, "Scavenging is a fine art. Not only must the cleaning be perfect, but the manner of doing it and the instruments used must be clean and not revolting to one's sanitary sense. If you become your own scavenger, not only will you insure perfect sanitation for yourself, but you will make your surroundings clean and relieve the scavengers of the weight of oppression which today crushes them. Do not imagine that thereby you would deprive them of their living. Today we have reduced them to the level of the beast. They earn a few coppers, but only at the expense of their human dignity. The same scavenger serves in the municipality as well as in your bungalow, with the result that he can do justice to neither. Look at him as he eats his food, coming under the shadow of the latrine wall, surrounded by filth. It is enough to break one's heart. It should not be difficult for you to find a more decent avocation for him to follow".

What qualities should such an honoured servant of society exemplify in his person? In Gandhiji's opinion, an ideal scavenger should have a thorough knowledge of the principles of sanitation. He should know how a right kind of latrine is constructed, and the correct way of cleaning it. He should know how to overcome and destroy the odour of excreta and the various disinfectants to render them innocuous. He should likewise know the process of converting nightsoil and urine into manure. But that is not all; Gandhi's ideal scavenger would know the quality of nightsoil and urine. He would keep a close watch on these and give a timely notice of the results of his examination of the excreta. That presupposes a scientific knowledge of the requirements of his profession. He would likewise be an authority on the subject of disposal of nightsoil in small villages as well as in big cities and his advice and guidance in the matter would be sought for and freely given to society. It goes without saying that he would have the usual learning necessary for reaching the standard here laid down for his profession. Such an ideal Bhangi*, while deriving his livelihood from his occupation, would approach it only as a sacred duty. In other words, he would not dream of amassing wealth out of it. He would consider himself responsible for the proper removal and disposal of all the dirt and nightsoil within the area, which he serves and regards the maintenance of healthy and sanitary conditions within the same as the summum bonum of his existence.

The views of Gandhiji influenced the state action soon after independence. The amelioration process started crafted within the overall framework of the constitution with careful provisions for the advancement of the socially, educationally and economically disadvantaged sections of society. Since the scavengers were the lowest stratum in society and were treated as untouchables amongst the untouchables, special schemes for their liberation and rehabilitation were introduced.

STATE ACTION FOR LIBERATION OF SCAVENGERS

The Constitution of India provides Fundamental Rights to the citizens of the country, in which Article (15) emphasises 'prohibition of discrimination on grounds of religion, race, caste, sex or place of birth', and its sub-clause (4) explains that nothing in this Article 'shall prevent the State for making special provision for the advancement of any socially and educationally backward classes of citizens or for scheduled castes and scheduled tribes.' Further, the Directive Principles of State Policy in Article (46), advocate for 'promotion of educational and economic interests of the scheduled castes, the scheduled tribes and other weaker sections of the people' such as scavengers and sweepers etc. It clearly states that 'the State shall protect these castes from social injustice and from all forms of exploitation.'

In order to make constitutional provisions effective, a number of committees and commissions had been constituted by the Governments, both Central and States, from time-to-time, for the welfare of scavengers' community. The suggestions and recommendations of some of the main committees are referred below:

Backward Classes Commission (1953) : The first Backward Classes Commission was constituted by the Ministry of Home Affairs, Government of India in 1953 under the Chairmanship of Kaka Saheb Kalelkar, which submitted its report in 1955. The Commission described the conditions of scavengers as sub-human. The Committee observed that "any society that expects one section of humanity to perform such humiliating and degrading service should be ashamed of itself". The Commission suggested that the hand collection and carriage of nightsoil should be replaced by mechanical cleaning of latrines with the provisions of sewers, drains etc. by the local authorities. It also emphasised that training in cleanliness and hygiene should be imparted to scavengers. Further, the report observed that if the scavengers were provided better housing and infrastructure facilities in their localities that would improve their living standards and help in their absorption in the society. The observations and the recommendation of the Commission were brought to the notice of all the State Governments by the Ministry of Home Affairs in October 1956 emphasising the need to introduce mechanical and up-to-date methods of cleaning latrines so that the inhuman practice of doing this work was obviated.

Protection* of *Civil Rights (1955): The Action by the Government of India was initiated by enacting a legislation known as Protection of Civil Right (PCR) Act, 1955. This Act became the base for conŕsidering implementation of different schemes by the Central and State Governments for Scheduled Castes, Scheduled Tribes and specially vulnerable groups of Society, such as scavengers, sweepers, bonded labourers, and nomadic and denotified communities for their social and economic development. The Act stresses on the formulation of effective schemes for the uplift and developŕment in terms of education, housing etc. after studying specific and composite problems of these communities to solve them effecŕtively.

The *Scavenging Conditions Enquiry Committee (Prof. N.R. Malkani Committee) (1957)* : The Malkani Committee was first of its kind which was set up on October 12, 1957 in pursuance of the recommendation of the Central Advisory Board for Harijan Welfare with a view to formulating "a scheme for putting an end to the practice of carrying nightsoil in baskets and buckets as head-loads." The committee made comprehensive study of the problem and submitted its report in 1960. It recognised that "the best system for disposal of nightsoil through mechanised means is the flushout latrines, but unfortunately introduction of this requires lot of funds and seems outside the scope of practicability in the foreseeable future. The question, however, arises of providing alternative methods of disposal of nightsoil taking into consideration the magnitude of the problem." The major recommendations of Malkani Committee were concerned with the improvement of working and service conditions of scavengers by introduction of mechanical tools such as wheel barrows, scrappers, rubber gloves and boots etc. as well as construction of houses and infrastructure facilities like water supply, electricity in their localities. Besides, it suggested that the scavengers should be paid wages according to Minimum Wages Act and other allowances, benefits and condition of work should be given to them also by the local authorities. The report of the Committee was informed by a genuine feeling for the scavengers' lot and contained several concrete recommendations not merely for eliminating the practice of carrying nightsoil as headload but also for improving the working and living conditions of the scavengers. The Committee had observed that "as long as dry latrines continue, the problem will exist". The recommendation and observations of the committee were sent to all the State Governments by the Ministry of Home Affairs for implementation. A Centrally Sponsored Scheme for improving the working and living conditions of sweepers and scavengers was introduced during the Third Five Year Plan. The scheme could not make much headway due to its several financial and technical shortcomings and was therefore discontinued during the Fifth Five Year Plan.

Committee on Customary Rights to Scavenging (1965) : The Department of Social Welfare, Government of India appointed in 1965 a committee to examine the question of abolition of customary rights of scavengers under the chairmanship of Prof. N. R. Malkani. The Committee submitted its report to the Government in 1966. The Committee found that when scavenging was not municipalized the latrines were cleaned privately and one particular scavenger acquired hereditary rights to clean such latrine as against another scavenger by an understanding and agreement. A customary relationship also developed with the householder and the scavenger received payments in some form or the other. The response of the Governments towards various recommendations of the committee was so lukewarm that no action could be taken.

National Commission on Labour (1968) **:** The Commission also studied the working and service conditions of sweepers and scavrengers for the enforcement of Minimum Wages Act. The report observed:

"The question of working conditions of sweepers and scavŕengers is not merely a matter of employer/employee relationŕship, but an issue that has figured prominently in the social and political discussions of the country for almost a generaŕtion. The question of untouchability, caste differentiation, the stigma and repugnance attached to this occupation, the ageŕ-long suppression, privations and affections suffered by these communities of sweepers and scavengers involve the social conscience of the whole nation. The uplift of this class of people requires breaking of traditional attitudes and a long process of education not only of the higher castes but of sweepers and scavengers themselves, to pull them up from the state of despair to which they have a fatalistic approach".

The report further observed:

"That a section of humanity should be condemned to such degrading condition is a slur on our culture and a disgrace to our society."

TASK FORCE FOR TACKLING THE PROBLEMS OF SCAVENGERS

In July 1989, the Planning Commission constituted a Task Force for suggesting measures to abolish scavenging with particular emphasis on rehabilitation of scavengers. The terms of reference of the Task Force were as follows:

i) To review and assess the problem and dimension of scavenging in terms of population and families engaged in the occupation, the number of towns and other habitations where it is practiced;

ii) To analyse the full magnitude of the problem with a view to suggesting a time-bound comprehensive programme and cost estimate for abolishing this practice along with advance steps necessary for training and rehabilitation of people now engaged in the profession;

iii) The type of intervention including legal measures required;

iv) Cost-effective technology for conversion of dry latrines, particularly in already built up old cities/ towns;

v) To identify agencies to deal with this problem in a well coordinated manner.

The report of the Task Force was published by the Planning Commission in March 1991. Earlier reports of committees and commissions on tackling the problems of scavenging had emphasized how to make the manual handling of human excreta easy and less hazardous, besides improving the living conditions of the scavengers. However, the reports did not point out that the core problem lay in the conversion of dry latrines to pourflush latrines, which alone could help eradicating scavenging from the very root. This was clearly brought out by the Task Force suggesting that the conversion be taken up in a phased manner.

The task force made four important recommendations:

i) A time-bound programme may be launched to convert all dry latrines in the country by the end of the Eighth Plan;

ii) Eradication of scavenging to be achieved during the Eighth Plan with an annual outlay of Rs. 110 crores by the Government of India;

iii) The practice of carrying of nightsoil as head-load and its manual handling to be banned by law; and

iv) Legislation so enacted should prohibit construction of new dry latrines in the existing or future buildings.

Other no less important recommendations were:

i) The core problem of converting dry latrines into pourflush latrines be tackled in a systematic manner;

ii) The ad-hoc assistance provided by the Ministry of Welfare for enhancement of scholarship rates and coverage of prematric scholarship scheme should also be extended to day scholars;

iii) For rehabilitating displaced scavengers, suitable bankable schemes will have to be formulated with the component of bank loans, margin money loan and/or subsidy;

iv) The displaced scavengers or their dependents can be allotted shops in the markets by the municipalities where these persons can start business/trades with financial help from the Government of India and other agencies;

v) The municipal scavengers may be provided with suitable accommodation and educational facilities for their children;

vi) Proper schemes for training of displaced scavengers and their dependents in various skills and trades should be developed by State/Local authorities;

vii) There should be a separate Project Management unit in the Ministry of Welfare manned by adequate staff for implementation and monitoring of the programme.

viii) The byelaws drafted and circulated by HUDCO for regulating the construction, operation, maintenance and control of latrines have got to be adopted by urban local bodies as early as possible;

ix) A programme may be launched to convert all dry latrines in the country by the end of 1995 i.e. by the end of the Eighth Plan period;

x) Incentive of subsidy/loan should be made available to such households in urban or rural areas, who would mostly be from poorer sections;

xi) Urban local bodies should give top priority in allotting shops, giving licenses/permits for trade/ business etc. to the vocationally displaced due to abolition/shrinking of manual scavenging;

xii) Voluntary organisations may be encouraged to take up programmes for conversion of dry latrines and training of displaced scavengers.

CONCERN BY NATIONAL LEADERS

In sheer anguish, Mahatma Gandhi had said "I may not be born again, but if it happens, I will like to be born in a family of scavengers so that I may relieve them of the inhuman, unhealthy and hateful practice of carrying night soil".

Prime Minister of India, the late Shri Jawaharlal Nehru, mentioned that "the day everyone of our countrymen and women gets to use a toilet, I shall know that our country has reached the pinnacle of progress".

Late Smt. Indira Gandhi, Prime Minister of India, made a historic statement while addressing the World Health Organisation in Geneva on the Water Decade (1981-90). She said, "a distinguishing characteristic of an advanced country is the provision of clean drinking water to its people in urban and rural areas. Economically backward societies most of which are in the tropical or desert zones have a high incidence of water borne disease. But they do not have the large financial, technological and managerial outlay which modern sanitation demands. Their people also lack the education and the means to change their personal habits. Today mankind as a whole has the knowledge and the means to ensure basic sanitation all over the world. International cooperation can and must supplement the inadequate resources of poor countries to ensure the supply of clean drinking water and improve the arrangement of sewage disposal."

While appreciating the need for abolishing the inhuman system of carrying and handling human excreta manually, the late Prime Minister Smt. Indira Gandhi had observed in parliament in March 1983, "everybody is conscious of this fact, and we have discussed this many times and there is no doubt that this is a disgrace and it should be eradicated as early as possible. You cannot ban a thing unless you provide an alternative thing for them".

Being greatly concerned with this inhuman practice, the then Prime Minister of India, Late Sri Rajiv Gandhi wrote in February, 1987 to all the State Chief Ministers and Lt. Governors of the Union Territories as follows:

"The continued practice of scavenging is the most undesirable form of social derogation of a section of our society. Under the 20-Point Programme, 1986, we have determined to eradicate it. To do so, we will require the active cooperation of all citizens and particularly of local bodies and governments which are responsible for approving construction activity or have their own housing programmes. This will also require rehabilitation of scavengers which will mean training and re-settlement in new occupations. It is requested that you may kindly give this matter your personal time and attention so as to improve awareness and ensure the willing cooperation of all concerned.

There should be a firm decision not to construct any government buildings or aided buildings which will involve scavenging. Existing buildings should be converted to provide for low cost or water borne sanitation within a time frame. I am told that feasible models already exist and can be easily adopted. Municipal bodies should not be permitted to allow any fresh constructions without such facilities. For rehabilitation of scavengers, adequate funds are already available under the Special Central Assistance for the Scheduled Castes. The fullest use may kindly be made of these and other resources made available by the Centre and State Governments."

NATIONAL SCHEME OF LIBERATION AND REHABILITATION OF SCAVENGERS AND THEIR DEPENDENTS

The Government of India had launched this Scheme in 1980-81. It aimed at the total elimination of the inhuman practice of cleaning and carrying night-soil, conversion of existing dry latrines by constructing low-cost two pit pour flush units with appropriate variation, to suit local conditions; and liberation, training and rehabilitation of scavengers and their dependents in alternative gainful and dignified occupations.

The salient features of the Scheme are as under:

(a) The scheme provides support to scavengers belonging to both Scheduled Castes and non-Scheduled Castes communities.

(b) It is envisaged to adopt cluster (Basti) approach.

(c) Selection of beneficiaries for training and rehabilitation is done simultaneously with that of conversion of dry latrines serviced by the scavengers.

(d) Training of beneficiaries is arranged in accordance with the norms of the Ministry of Rural Development. The selection of trades is carried out after detailed counseling with the beneficiaries keeping in view their aptitude, preference and competence for particular trades. Private and public sector firms are contacted for this purpose.

(e) Project cost per beneficiary is Rs. 50,000. Group

projects are also allowed. Women scavengers are given preference.

(f) Efforts are made to ensure convergence of various development programmes operated by various ministries.

(g) In order to facilitate bank loans, the Reserve Bank of India issued a circular on March 20, 1993 stating that security for the loans will be only hypothecation of the assets created out of the loans/subsidy in favour of the bank. All loans upto Rs.6,500 are treated as loan under Differential Rate of Interest and concessional rate of interest of 4 per cent is charged.

The Scheme was started under the aegis of Ministry of Home Affairs and was later operated through the Ministry of Welfare, (now the Ministry of Social Justice and Empowerment). The scheme continued to be implemented through the Ministry of Welfare with certain modifications till the year 1990-91 when the Scheme was bifurcated and the programme of conversion of dry latrines into water-borne flush latrines was entrusted to the Ministry of Urban Affairs and Employment and the programme of training and rehabilitation of scavengers continued to be operated by the Ministry of Welfare. Accordingly the new scheme namely, 'National Scheme of Liberation and Rehabilitation of Scavengers and their Dependents was launched by the Ministry of Welfare in the year 1991-92.

The progress in regard to the training and rehabilitation of scavengers had been rather tardy in most of the States/ Union Territories. Although the guidelines were modified w.e.f. 1st April 1996 but the requisite results had not been achieved. Out of more than 7 lakh scavengers identified till them, about 2.50 lakh scavengers only had been rehabilitated. Under the Low-Cost Sanitation Scheme, out of the total number of 20,48,461 units for conversion, 15,35,562 units for new construction, and 3,463 units for construction as community toilets, the State Governments, as on 31st December 1997, had been able to complete only 7,22,810 and about 82,021 units were in progress for conversion. While the target was to convert all the dry-latrines and offer training and rehabilitation to all the scavengers and their dependents during the 8th Plan period, not even 50% of the target had been achieved in both the areas.

Various factors had been identified to be responsible for the poor performance which, inter alia, include : (1) inadequate stipend for training; (2) unviable projects taken up for the purpose of rehabilitation; (3) reluctance of banks to finance projects under the scheme; (4) weak and inadequate implementation machinery at the State level; (5) lack of adequate monitoring at the local and State level due to the ineffective functioning of the District and State Coordination Committees set up in several States; and (6) lack of effective coordinated implementation by the States of the schemes of the Ministry of Urban Affairs and Employment and the Ministry of Welfare.

The working group on Water Supply and Sanitation for the 9th Five Year Plan in their report identified the following reasons for poor performance of Low-Cost Sanitation Scheme for Liberation of Scavengers being implemented by Ministry of Urban Affairs and Employment.

- Lack of sufficient number of claims from the State Government;
- Delays in preparing document papers required for sanctioning of subsidy and loan for each individual beneficiary;
- Inability of the implementing agencies to draw the full amount of subsidy component due to the slow physical progress;
- Non-adoption of proper technology keeping in view the social needs;

- Lack of public awareness; and
- Reluctance on the part of the State Governments in giving guarantees.

These reasons figured prominently during the discussions held in series of meetings with the State Secretaries/Heads of local self-government department, representatives of the Ministry of Welfare and the Planning Commission.

After having closely monitored the progress of the Scheme, the Government of India, during the year 1990-91, bifurcated the Scheme, and the programme of conversion of dry latrines into water-borne flush was entrusted to the Ministry of Urban Affairs and Employment (now Ministry of Urban Development and Poverty Alleviation) and the programme of the Liberation and Rehabilitation of Scavengers and their dependents to the Ministry of Welfare (now Ministry of Social Justice and Empowerment). The scheme was launched by the Government of India on 22nd March 1992. After a review of the implementation of the Scheme, for its effective implementation, these modifications took effect from 1st April 1996.

The main components of the revised Scheme are as follows :

- Conversion of all existing dry latrines into water borne toilets.
- Training in identified trades for scavengers and their dependents at nearby training centres, keeping in view their aptitude and local requirement and environment
- Rehabilitation of scavengers in various trades and occupations.
- Placement services
- Client services like help in procuring loan, land, raw materials and marketing etc.
- Housing

- Sanitation and environment
- Public awareness and project preparation
- All such and related activities which are in harmony with the objectives listed above.
- Evaluation on continuous basis.

The effective liberation and rehabilitation of scavengers and their dependents has been proposed to be implemented through integrated comprehensive package approach on the whole town basis. Under the revised scheme, preference has to be given to rehabilitation of women, and for that, the specific women oriented schemes have to be formulated by the implementing agency. Similarly, the scheme has accorded priority to the rehabilitation of disabled scavengers.

The scheme as per the revised guidelines, has to be implemented through the registered NGOs, public trusts, charitable companies or other public bodies or institutions having a legal status, in addition to the implementation of the scheme through municipal corporation, municipality, local bodies and HUDCO. The training and rehabilitation of scavengers is to be implemented by the same organization in order to avoid any gap between the two and to have an integrated approach. Under the pattern of funding by way of subsidy etc. it has been proposed that subsidy worth Rs. 25,000/- shall be reimbursed under the scheme or a loan upto the ceiling of Rs. 50,000/- to be sanctioned by NSKFDC as against the existing provision of Rs. 7,500/- (margin money) and Rs. 10,000 (subsidy) for the loans granted by SCDCs upto the ceiling of Rs. 50,000/-.

To ensure that adequate public opinion is created against the inhuman practice of manual handling of human excreta, the implementing agencies and State Governments have been allowed expenditure upto 5% of the project cost. Detail mechanism to sanction projects and monitor the progress of implementation has been provided at the Central, State, District and Local body level. In order to strengthen the Scheme on a continuous basis, a provision of 2% of the allocation has been allowed to be earmarked for monitoring and evaluation by the expert evaluators and academic institution.

The Scheme envisages conversion of existing dry latrine into low cost water seal pour flush latrines and construction of new sanitary units where none exist to prevent open defecation. The basic objective is to liberate the scavengers from the obnoxious practice of carrying head loads of night soil and the scheme is being implemented on a whole town basis.

The scheme is being operated by providing subsidy from this Ministry and loans from HUDCO in a synchronized manner so that the conversion/construction of low cost sanitation units and liberation of scavengers is done effectively. The financing pattern in the programme is as under :

(1) For substructure i.e. upto plinth level :

Category	**Subsidy**	**Loan**	**Beneficiary contribution**
EWS	45%	50%	5%
LIG	25%	60%	15%
MIG/HIG	Nil	75%	25%

(2) For superstructure i.e. above plinth level :

Category	**Subsidy**	**Loan**	**Beneficiary contribution**
EWS	Nil	95%	5%
LIG	Nil	85%	15%
MIG	Nil	75%	25%
HIG	Nil	60%	40%

The HUDCO is extending loan at an interest rate of 8% floating or 8.5% fixed per annum. The loan is recoverable over a period of 7 years. The progress of the LCS Scheme in the country upto 31.03.2005 was as under :

1.	Total No. of schemes sanctioned by HUDCO	: 867
2.	No. of Towns covered	: 1534
3.	Project cost of sanctioned schemes	: Rs. 1,91,557.50 lakhs
4.	Subsidy sanctioned	: Rs. 53,628.58 lakhs
5.	Loan sanctioned	: Rs. 70,912.30 lakhs
6.	Subsidy released by HUDCO to State	: Rs. 31,197.65 lakhs
7.	Loan released by HUDCO to State	: Rs. 49,703.41 lakhs
8.	No. of units sanctioned for Conversion + Construction	: 49,12,195
9.	No. of units completed (Conversion + Construction)	: 19,16,510
10.	No. of units in progress (Conversion + Construction)	: 3,36,543
11.	No. of scavengers to be liberated with the implementation of sanctioned schemes	: 1,18,464
12.	No. of scavengers liberated	: 41,559
13.	No. of towns declared scavenger-free	: 586

The scheme has been able to convert/construct so far only 45* percent of the total number of dry latrines in the urban areas. The main problem faced by HUDCO in the implementation of this scheme are :

---

*Evaluation and Impact Assessment of ILCS AFC – Oct. 2007.

(i) Non-submission of utilization certificate by the States.

(ii) Lack of new schemes/proposals sent by the State / Local Bodies

(iii) Unwillingness among the beneficiaries to bear the beneficiary contribution and subsequent repayment of loan

(iv) Delay in giving guarantee by the State Governments.

EMPLOYMENT OF MANUAL SCAVENGERS AND CONSTRUCTION OF DRY LATRINES (PROHIBITION) ACT, 1993

Conversion of dry latrines numbering about ten million in the country is a mammoth task. Given the slow progress in regard to the conversion of dry latrines into pour-flush latrines, the country would take several decades to convert all dry latrines. The existence of dry latrines is the root cause of manual scavenging and therefore, the country would have to wait for eradication of scavenging until a time when there wouldn't be dry latrines which require the services of scavengers. As the shame of manual scavenging is becoming increasingly embarrassing, the Central Government decided to bring in the element of coercion in its approach to solving the problem through a legal measure. It was considered necessary to go in for the desired abolition of manual scavenging by enacting a legislation – Employment of Manual Scavengers and Construction of Dry Latrines (Prohibition) Act, 1993, by taking recourse to the provisions of Articles 249 and 250 of the Constitution of India. The legislation has been enacted out of the realisation that municipal laws by themselves are not stringent enough as a measure for conversion of dry latrines into water-seal latrines, as also for the prevention of construction of dry latrines. In view of this it was found necessary to enact a uniform legislation for the whole of India for abolishing manual scavenging by declaring employment of manual scavengers for removal of human excreta an offence and thereby ban the future proliferation of dry latrines in the country. The legislation was considered desirable for eliminating the dehumanising practice of employment of manual scavengers and for protecting and improving the human environment to make it obligatory to convert dry latrines into water-seal latrines or to construct waterseal latrines in new constructions. The Act provides that no person shall engage in or employ for or permit to be engaged in or employed for any other person for manually carrying human excreta; or construct or maintain a dry latrine. Section 14 of the Act provides that "Whoever fails to comply with or contravenes any of the provisions of this Act, or the rules or schemes made or orders or directions issued there under in respect of each such failure or contravention shall be punishable with imprisonment for a term which may exceed to one year or with fine which may extend to two thousand rupees, or with both, and in case the failure or contravention continues, with additional fine which may extend to one hundred rupees for every day during which such failure or contravention continues after the conviction for the first such failure or contribution". Section 15 of the Act provides that nothing contained in Section 14 shall render any such person liable to any punishment, if he proves that the offence was committed without his knowledge, or that he has exercised all due diligence to prevent the commission of such offence.

Notwithstanding the extensive powers conferred on the State Governments to prevent maintenance and construction of dry latrines and engagement of manual scavengers, which has become penal offence in areas and from the date notified by the concerned State Governments, the conduct of the Central and the State Governments in relation to the implementation of the Act compels one to come to a conclusion that the Act is nothing better than a dead letter word (Yashwant Raj 1997 : 73).

Some of the important reasons listed by Yashwant Raj (1997) for the poor implementation of the Act are :

- The Act was passed on 3rd May 1993 and was published in the Official Gazette on 7th June, 1993. But the Central Government took more than three and half years to issue the necessary notification specifying the date on which the Act shall come into force in six states (Andhra, Goa, Karnataka, Maharashtra, Tripura and West Bengal and in all UTs, which had consented to the Central Government, enacting a legislation for them, notwithstanding the fact that the subject matter of the Act is in the State list.
- Even in these six states the Act did not come into force for full three and a half years because the Ministry of Urban Affairs and Employment could not issue necessary notification, and therefore, the State Governments could not frame rules and appoint designated authorities to operationalise different provisions of the Act.
- The Act had limited outreach (only six states and UTs); the four States, viz., Haryana, Orissa, Assam and Punjab adopted the Act but have not enforced it. And, the States of Gujarat, Rajasthan and Bihar have only given assurances to adopt it. The rest of the States have shown little enthusiasm in the matter. They have either not replied to the request or have sent evasive replies.
- The Act, which has the potential of becoming powerful tool in the hands of the State Governments in their efforts to banish manual scavenging has so far proved to be a damp squib largely due to indifferent attitude of the Central Government and State Governments. The implementation of the Act is a sad commentary on the sincerity of the Governments in regard to the objective of abolition

of manual scavenging and exemplifies the ever existing difference between words and deeds of the Governments.

NATIONAL COMMISSION FOR SAFAI KARMACHARIES ACT, 1993

Considering the persistent demand of various associations of scavengers regarding the need for constituting a National level statutory Commission, on the pattern of the National Commission for Schedule Castes and Scheduled Tribes, the National Commission for Minorities and the National Commission for Backward Class, the Central Government enacted the National Commission for Safai Karmacharis Act, 1993 and constituted the Commission on 14.8.94. The functions and responsibilities attached to the Commission are fairly comprehensive and wide ranging. Section 8 of the Act, prescribes the functions of the Commission, as below :

(a) To recommend to the Central Government specific programmes of action towards elimination of inequalities in status, facilities and opportunities for Safai Karmacharis under a time-bound action plan;

(b) To study and evaluate the implementation of the programmes and schemes relating to the social and economic rehabilitation of Safai Karmacharis and make recommendations to the Central Government and State Governments for better co-ordination and implementation of such programmes and schemes;

(c) To investigate specific grievances and take suo moto notice of matters relating to non-implementation of : (i) programmes or schemes in respect of any group of Safai Karmacharis; (ii) decision, guidelines or instruction, aimed at mitigating the hardship of Safai Karmacharis; (iii) measures for the social and economic uplift of

Safai Karmacharis; (iv) the provisions of any law in its application of Safai Karmacharis; and (v) take up all relevant matters with the concerned authorities or with the Central or State Governments.

THE PROFILE OF STATE ACTION

After considering the recommendations of various committees and commissions, a Centrally-sponsored scheme, "Improvement of Working and Living Conditions of persons engaged in unclean occupations" was introduced during Third and Fourth Five Year Plans, which provided Rs. 1.44 crore and Rs. 3.00 crore respectively. But the scheme did not contribute to decreasing this demeaning practice due to failure of the State Governments to introduce organisational frame and financial support to the local bodies. Thus, the initial action programme initiated by the Ministry of Works and Housing, Government of India for Liberation of Scavengers did not make reasonable impact and was dropped because the funds (Rs. 4.4 crore) provided in the Fifth Five Year Plan for this scheme were inadequate which covered only a small number of towns.

It was, however, at the beginning of the Sixth Five Year Plan, that Centrally-sponsored Scheme of Liberation of Scavengers was sincerely introduced by the Ministry of Home Affairs, Government of India under Protection of Civil Rights Act, 1955. For accelerating the process of conversion of dry-latrines into pour-flush latrines for liberating scavengers and their dependents, some circular letters were issued to all the State Governments and Union Territories by the Ministry of Home Affairs from June 15, 1981 to June 21, 1985 emphasising therein (a) conversion of all the existing household/public dry-latrines into pour-flush latrines on the whole town approach basis in selected small and medium towns; and (b) rehabilitation of unemployed scavengers in alternative employment/occupation simultaneously. The said Ministry while issuing a circular letter No. BC-14011/

25/85 PCR-CELL, dated June 21, 1985 further emphasised the urgent need for implementation of this scheme along with the detailed guidelines of the scheme.

Subsequently, the Scheme of Liberation of Scavengers was transferred from the Ministry of Home Affairs to the Ministry of Welfare, Government of India in 1985. And, in continuation of the initiative taken earlier for this scheme, the Ministry of Welfare laid down two components of the Scheme in its circular letter No. DO.BC. 11014/1/86-SCBCD:III dated January 20/22, 1986 as follows:

i) conversion of dry-latrines into water borne/sanitary ones in selected small and medium towns on the 'whole town' approach basis – to be implemented by the Urban Development/Local Self-government of the State Governments; and

ii) providing necessary training to the scavengers and their dependents - follow up action for their occupational mobility by the welfare development or the nodal department for overall development of scheduled castes/ scheduled tribes and specially vulnerable groups such as scavengers, sweepers and denotified communities.

Further, the Ministry of Welfare issued another circular letter dated May 26, 1987 for making best use of central assistance for speedy implementation of PCR Act and Liberation of Scavengers Scheme by the State Governments and Union Territories.

The movement of liberation of scavengers through conversion of dry-latrines was actually initiated by some State Governments through voluntary organisations namely Harijan Sevak Sangh in Gujarat, Sulabh International in Bihar and Gandhi Smarak Nidhi in Maharashtra in sixties and early seventies. These States passed legislations with regard to conversion and liberation of scavengers. The implementation of these schemes, however, was taken up in a massive way in 1980-81 at All India level by the Ministry of Welfare, Government of India, 25 years after passing the legislation on Protection of Civil Rights Act, 1955. Concrete action for implementation was initiated by the said Ministry, when it was decided to meet 50 per cent of the cost of conversion of dry latrines from the Central Sector as grants-in-aid, while the rest would be provided by the State Governments as matching share in the form of loan to the beneficiaries of P.F. latrines to be recoverable in easy instalments. The Ministry of Welfare released Central assistance of Rs. 82.05 crore upto March 1990 to the State Governments in respect of 490 towns.

Besides the Ministry of Welfare, the implementation of the scheme of Low-cost Sanitation is being implemented by Ministry of Environment under Ganga Action Authority, Ministry of Urban Development under the Integrated Development of Small and Medium Towns, HUDCO, Ministry of Rural Development as well as by voluntary organisations viz. Sulabh International operating in many States, and Safai Vidyalaya in Gujarat. It is reported that rehabilitation of liberated scavengers as a result of conversion of dry-latrines has not received due attention of any Ministry/Departments, which are implementing the scheme of Low Cost Sanitation, other than the Ministry of Welfare and Voluntary Organisations like Sulabh International and Safai Vidyalaya, the reason being that the Ministry of Welfare and Voluntary Organisations have undertaken the scheme of Liberation of Scavengers from the social and humanitarian point of view whereas other organisations have taken it up from sanitation point of view only. Hence, there is a lack of co-ordination in implementation strategy in both the schemes of Low Cost Sanitation and Liberation of Scavengers. Unless there is an effective co-ordination at the macro level in terms of organisational, financial and management control through scientific information system, the problems of conversion of dry-latrines as well as Liberation and Rehabilitation of displaced Scavengers and their dependents at the micro level through institutional arrangement for monitoring and periodic evaluations of both the schemes could not be resolved.

For resolving the above issues, a Coordination Committee had been set up at the Central level with HUDCO as a convenor and the representatives of the Ministry of Urban Development, Ministry of Welfare, Ministry of Environment and Forests, the Planning Commission and the Central Ganga Authority as members. Reprersentatives of voluntary organisations viz. Sulabh International and Safai Vidyalaya were also invited for consultations. The recommenrdations of this Committee are as follows:

(A) ***Financial*:** The financial pattern of low-cost sanitation scheme has been modified as per recommendations of the Coorrdination Committee from the financial year (1990-91) as indicated below:

   a) Economically Weaker Sections (EWS): 45 per cent subrsidy + 50 per cent Loan + 5 per cent beneficiary contrirbution.

   b) Low Income Group (LIG): 25 per cent subsidy + 60 per cent Loan + 15 per cent beneficiary contribution.

   c) MIG/HIG: Nil subsidy + 75 per cent Loan + 25 per cent beneficiary contribution.

(B) ***Rehabilitation* of *Displaced* Scavengers** : (i) The Ministry of Welfare is concerned with the welfare of scheduled castes/schedruled tribes and vulnerable groups such as scavengers, sweepers and denotified communities. The work of rehabilitation has been entrusted by this Ministry to the Department of Welfare in the State Governments. The Scheduled Castes Development Corporations (SCDC) play a major role in rehabilitating the unemployed scavrengers in alternative jobs. The municipal scavengers have been absorbed by the local authorities in alternative jobs after conversion of dry-latrines in a number of towns and cities. The problem arises in case of their dependents (children and private scavengers). For this the Ministry of Welfare has authorised the State Governments to utilise 10 per cent of the Special Central Assistance (SCA) to Special Component Plan (SCP) for scheduled castes and rehabilitation of vulnerable groups, including scavengers. The SCA is provided to the State Governments on 100 per cent basis. The State SCDC have to formulate suitable schemes for rehabilitation of unemployed scavengers or their dependents in different occupations and trades. Financial assistance is arranged and provided to the beneficiaries through margin money loan, subsidy and bank loans. The margin money loans are provided to State SCDC by the Central and State Governments in the ratio 49:51.

(C) ***Training of* Scavengers and the Dependents/Wards :** The Ministry of Welfare have advised the State Governments to initiate suitable training programmes for scavengers, and their dependents/wards for the suitable occupational/job opportunities. In this regard, the Government of Bihar initiated the first training centre of scavengers in the country which was set up by the Sulabh International as Sulabh Institute of Research and Rehabilitation at Patna. This Institute imparted training to 2,564 scavengers and their dependents/children for three years from 1985-86 to 1987-88 in nine trades such as tailoring, mechanics, typing, driving, manufacture of cane materials and leather goods, pan manufacturing, carpentry and electrician. The training was the responsibility of the Sulabh Institute, whereas rehabilitation of the trained scavengers' children was the responsibility of the Bihar Scheduled Castes Development Corporation.

> The findings of the recent survey on employment position of the scavengers' children conducted by the Council for Social Development in November 1991 indicated that 61.6 per cent of the children were unemployed, 26.4 per cent were employed and 12.0 per cent were self-employed. It seems that Bihar Scheduled Castes Development Corporation took no initiative to assist the trained children in settling in different occupations and trades, through various rehabilitation schemes. Moreover, for non-release of funds by Bihar Government in the fourth session (1988-89), the Sulabh Institute which imparted training to these children was closed. However, such experiments are being carried in Maharashtra (Jambhol-60 km from Bombay) and Rajasthan also, where vocational training centres have been set up on the lines of the Sulabh Institute of Research and Rehabilitation. As reported, the other State Governments have taken initiative to provide training to liberated scavengers' dependents/ children through existing vocational training centres such as Industrial Training Institutes etc. Training, however, should be considered as a means to achieve an end for rehabilitation of these children in an effective way. This calls for Integrated Action by different institutions/departments of the State Governments and the people concerned in the context of liberation of scavengers as a whole.

While there is no denying the fact that much significant work has been done in the post-Independent India to liberate scavengers from the social bondage, the problem without doubt, persists as a continuing disgrace for the nation. Many welfare schemes launched for the liberation of scavengers have not made any significant dent in the problem.

The evaluation studies made to assess the impact of these governmental programmes revealed that the efforts fell far short of what is really required to liberate the scavengers, mainly because of the absence of priority to the programme in the Government's planning process. The studies have revealed that despite the emphasis by all political leaders and efforts made by the Government since Independence and inclusion of eradication of scavenging in the revised 20-point programme of 1986, tangible improvement has not been achieved in a period of more than four decades. The extent of scavenging in the country perhaps stands at the same level or may have increased, with increase in the haphazard growth of urbanization and civic services having remained far behind.

The attitude of the scavengers themselves has been a strong deterrent in the conversion of latrines and their rehabilitation. This is primarily due to the following reasons:

1. Since scavenging has been a way of life for the family of the scavengers, they fear that their livelihood may be affected if bucket latrines are abolished.
2. The scavengers have a fatalistic outlook due to lack of education and proper guidance regarding the openings for alternate employment. On account of this they resist any change in the status quo.
3. Vast majority of sweepers and scavengers are unaware of various measures undertaken by the government in respect of financial assistance, training and employment reservation.

REVIEW OF THE MAJOR SCHEMES FOR SCAVENGERS

Despite widespread denunciation and disapproval of manual scavenging, this abhorrent evil persists in many parts of the country. Recommendations of successive committees have not been implemented to effectively tackle the problem. There is variety of reasons for that, most being socio-economic as well as psychological. Many scavengers who continue their hereditary occupation think that this is the only occupation that has earned them their living. Besides that, most such scavengers feel that they are not educationally, vocationally and attitudinally well-equipped to shift to other dignified occupations. The amazing part, however, is that many, if not most, do not see anything wrong in manual scavenging, which is like any other occupation; other people are engaged on strictly hereditary basis. Some of them even welcome it because there is no competition. All scavengers, however, do not like manual cleaning of human excreta, dirt and garbage. They want change in their occupational pattern, but are forced to stick to this abominable occupation for want of availability of alternative sources of employment. Interestingly, the female scavengers are more keen to get away from this sub-human occupation and are willing to take up alternative occupations for a more dignified life.

Regardless of the official assertions and averments regarding lot of good work being done to liberate scavengers from their hereditary occupations through the provisions of training and rehabilitation, the ground reality does not suggest that the targets are being energetically chased, programmes have been well-executed and the most needy amongst the scavengers have been provided with grants, concession and facilities. The result is that the schemes for their welfare and development, training and rehabilitation have only made a marginal impact on the socio-economic and working conditions of scavengers and their dependents. Less than one-quarter of the total population of the scavengers (4,18,633) as estimated by the Planning Commission's Task Force Report (March, 1991) have been liberated and rehabilitated. The National Commission for Safai Karmacharis is so sceptical of the official statistics which do window dressing and tend to show-case the progress of the schemes, that it dismissed the statistics released by Ministry of Welfare, saying that 18,583 scavengers were rehabilitated during 1992-93, 40,820 during 1993-94, 64,967 during 1994-95 and over 1,00,000 during 1995-96. The

Commission dismissed these statistics as cooked up and deceptive.

The Scheme for the Liberation and Rehabilitation of Scavengers and their Wards has been found to be both structurally and functionally flawed on number of counts. Whilst the money earmarked and spent by the State Governments is meagre in face of the complexity of the situation even the full amount of the meagre allocation has not reached the needy and deserving scavengers. Much of it has been gobbled up by the middlemen in collusion with corrupt officials and implementing agencies and banks. Furthermore, the selections of projects have been influenced by considerations other than their merit and viability. Given these implementation bottlenecks and the poor impact of the Scheme, it is difficult to describe the scheme as an instrument of social transformation, as it was conceived to be.

One major flaw of the Scheme is that it operates on an assumption that poverty drives the scavenger families to stick to their old hereditary occupation, and that the whole community of scavengers lives in penury. Because of this assumption the Scheme has all the ingredients of an income augmenting scheme. Several surveys on the economic conditions of the scavenger communities in many states have revealed that income-wise most scavengers are above the poverty line, and hence economic approach to solving a social problem of ancient vintage has its own limitations. Moreover, to believe that a project with seed money of 50 thousand will fetch the beneficiary enough income in urban areas to abandon his / her hereditary occupation is fallacious. Any good and financially viable income generating project these days would require far more money than what is provided under the Scheme. Another obvious flaw in the scheme is that it leaves the beneficiary of the sanctioned projects free form the obligation of being a real stakeholder. In the absence of any financial stake, the beneficiary remains uncommitted to make the project income generating and profitable. He / She thinks that even if the project does not do him /her any good, he/she had nothing to lose and can always revert to one's own old occupation.

In regard to the educational uplift of the wards of the scavenger, the present facility of providing stipend of Rs. 150/- per month has had its own set of problems. The amount is ridiculously small and can not be taken as good enough for the sons and daughters of the scavengers to continue attending school. There is therefore, a lamentable absence of formal education amongst the sons and daughters of scavengers. Therefore, to expect them to take up self employment projects without formal education and / or training is a pipedream.

The Scheme of Pre-Matric Scholarships for Children of those engaged in unclean occupations distributes only doles which have made little impact on raising the educational status of the sons and daughters of the scavengers. While the Scheme is commendable, the rate of scholarship is too meagre both for day scholars and hostellers to cope up with present-day inflationary trend.

PERSPECTIVE

The task of making the country free from scavenging entails enormous difficulties. It calls for good planning, administration, financing, implementation with back-up technology and follow-ŕup. All these aspects have to be carefully considered and the programmes devised accordingly. To be sure the planning for the liberation of scavengers should involve the following:

1. Estimating the total number of scavengers in the entire country who are presently engaged in the demeaning task of handling human excreta.
2. Estimating the number of scavengers already liberated from different cities and towns due to

conversion and / or construction of low-cost latrines and how they are employed presently.

3. Assessing the present number of dry/bucket latrines on town by town basis.
4. Setting target of liberation of scavengers by converting / constructing low-cost latrines.
5. Fixing programmes of training and rehabilitation of the liberated scavengers and their wards.
6. Identifying constraints and devising solutions to meet the constraints.
7. Setting investment priority in the plan, budget and idenŕtifying specific projects.

SOME CRITICAL NEEDS

In order to make the policy and programmes for the liberation of scavengers more effective, following are the critical needs:

1. There is a need to launch a better and effective promoŕtional and educational campaign both on social and health aspects.
2. There is a need to examine and resolve the weak infraśtructure and financial problems both at the Government and Local Authorities level.
3. There is need to revitalise the political will and policy commitment backed by decision making apparatus at the Centre, State and local level.
4. There is need to provide technical support with advanced technologies suited to different areas and situations.
5. There is need to study the financial aspects in terms of availability of funds and the capacity to bear cost and repayment of investment by the householders and the local body.
6. There is need to launch a strong movement for social

reforms to mould the attitude of scavengers to leave their age-old profession of scavenging and move into occupartions not connected with scavenging.

7. There is need to provide adequate facilities for training the liberated scavengers and their wards in different trades and skills and their rehabilitation.
8. There is need for training the manpower necessary for implementing the programme.
9. There is a need for making vigorous use of print and electronic media for the awareness campaign about the programme.

REFERENCES

Barve, B.N. (1958) The Scavengers' Living Conditions Enquiry Committee, Government of Maharashtra, Mumbai.

Fuchs, Stephen (1981) At the Bottom of Society : The Harijan and other Lower Castes, Munshi Ram Manohar Lal Publications, New Delhi.

Government of India (1992), National Scheme for Liberation and Rehabilitation of Scavengers and their Dependents.

Malkani, N.R. (1966), Scavenging Conditions – Enquiry Committee, Ministry of Home Affairs, New Delhi.

Malkani, N.R. (1969), Clean People and Unclean Country, Harijan Sewak Sangh, New Delhi.

Malkani, N.R. (1966), The Committee on the Customary Rights to Scavenging, Ministry of Home Affairs, Govt. of India, New Delhi.

Mukhopadhay K.K. (1993), Scavengers in Delhi: A Report, Department of Social Work, Delhi University, Delhi.

Pathak, Bindeshwar (1991), Road to Freedom – A Sociological Study on the Abolition of Scavenging in India, Motilal Banarsi Dass Publishers, Delhi.

Pathak, Bindeshwar (1981), Sulabh Shauchalaya : A Simple Idea That Worked, Amola Prakashan, Patna.

Pandhya B.P. (1969), Committee on Conditions of Sweepers and Scavengers, National Commission on Labour, New Delhi.

Planning Commission (1991), Report of the Task Force for Tackling the Problems of Scavengers and Suggesting Measures to Abolish Scavenging with Particular Emphasis on their Rehabilitation, New Delhi.

Salappa, IPD (1976) Committee on Improvement of Living and Working Conditions of Sweepers and Scavengers, Government of Karnataka.

Shyamlal (1992) The Bhangi : A Sweeper Caste – Its Socio-Economic Portraits, Popular Prakashan, Mumbai.

Sinha B.D. (1994), Evaluation Study on National Scheme of Liberation and Rehabilitation of Scavengers in Uttar Pradesh, Council for Social Development, New Delhi.

Yashwant Raj (1997), The ugly Face of Manual Scavenging in India. A Dissertation Submitted to the Indian Institute of Public Administration, New Delhi.

Chapter -V

SANITATION MOVEMENT : EVOLUTION AND GROWTH

Sanitation in India became a subject of public discourse in the context of Mahatma Gandhi's passionate advocacy of liberating scavengers from the age-old shackles of untouchability. He himself cleaned latrines and did scavenging work to impress upon the people that no work was polluted and no one could be regarded untouchable only because he/she did scavenging. At his Sabarmati Ashram, Gandhiji made scavenging as a pre-condition for joining the Ashram. Consequently, his followers who were from different castes, had to perform sanitation work and no scavenger was employed to do scavenging. In this way, the sanitation movement was launched by Gandhiji not primarily for keeping the Ashram clean, but also to do away with the traditionally established practice of hereditary occupation. Gandhiji was not opposed to scavenging but rejected the concept of superiority or inferiority associated with the work. To him all jobs and occupations were equally important. He maintained that scavenging, or sanitation work, could be taken up by anybody irrespective of his/her caste and creed. He opposed theocratisation of the job and stood for secularization and rationalization of occupations.

An evaluation of the sanitation work done by Gandhiji and his associates reveals that all of them took up sanitation programme in the context of scavenging. The main task before them was that of sweeping and cleaning latrines and engaging people from different castes and social background in that work. The sanitation work taken up in Sabarmati

Ashram by Gandhiji, or at 'Bhangi* Basti' by Vithal Phalke or the village sanitation programme by Shri Balcoba Bhave or the education of rural sanitation by Shri Jagat Ram and other such programme have been briefly described in the preceding chapter. It is evident from these programmes of action that sanitation work was started in the background of scavenging and with the basic objective of emancipating scavengers.

It is also evident from an analysis of earlier sanitation programmes that most of these efforts and struggles were confined to village sanitation. In the urban centres, there were municipalities and corporations to look after sanitation work, and that in many urban areas water-carriage system was adopted for the disposal of night-soil in some of the cities. But, in villages there were no such agencies, nor was the disposal of night-soil through water-carriage system possible due to shortage of water supply and absence of any agency to undertake this gigantic task. It was also realized by the followers of Gandhiji that the programme of rural sanitation had no meaning without adopting an adequate and proper system for the disposal of nightsoil. Usually, in village areas, there was no latrine system except in a few village communities. The people were used to defecate in the open either in the fields or at some other isolated places. The isolated places were used by adults, male or female, but the minors used to defecate in the open space within the locality. Given this reality the problem of environmental pollution was bound to arise due to inadequate system of disposing night-soil. Hence, the programme of village sanitation was given priority for checking environmental pollution. For this purpose, it was required to develop devices for disposing night-soil without manual scavenging being done by the persons belonging to scavenging communities/ castes.

Appa Saheb Patwardhan took a leading role in this regard. He knew that India was extremely poor and any high-cost programme of sanitation in villages would fail. Hence, he designed the low-cost sanitation programme and developed different types of low-cost latrines for the disposal of night-soil. The 'trench latrines' and 'mobile trench latrines' introduced by Appa Saheb, have been briefly described in earlier chapter. These latrines were designed to be used by villagers in the agricultural fields. Provision was also made for covering the seat from all the four sides to maintain privacy and, hence, both men and women could use this latrine at any time during day and night. A trench was dug under the seat of the latrine for deposit of night-soil. When it was filled in, it was fully covered by soil and another trench was dug at a short distance from the first and the seat and screen covering from four sides were shifted. Later, he designed mobile trench latrines and, for this purpose, four wheels were provided in the four pillars or stands of the seat. When one trench was filled in, another trench was dug and the seat along with the screen was moved over to the other trench with the help of the four wheels. In this way the mobile trench latrine helped people save time and money on the construction of new seat and on providing curtains. This type of latrine was very easy to design and the villagers had built many such latrines by the side of their fields. After some time the covered night-soil was decomposed and transformed into a good manure to be used in the fields. The problem of environmental pollution was also checked and there was no need to engage scavengers for cleaning the night-soil and carrying it to some other place.

The mobile trench latrines were also helpful in liberating scavengers, checking environmental pollution and in producing quality manure. In addition to it, Appa Saheb Patwardhan carried out a series of experiments at his Gopuri Ashram in Ratnagiri district of Maharashtra with the objective of designing cheaper and hygienic latrines for villagers. On the basis of his experiments, Appa Saheb developed two designs of low-cost latrines known as 'Sopa type' or 'Sopa Sandas' and 'Naigaon latrine'. In the 'Sopa type' latrine a leaching pit was provided for the disposal of night-soil; whereas in case of 'Naigaon latrine' arrangement was made to collect the night-soil in a container to be used as manure for agricultural operations. In designing both these latrines, Appa Saheb kept two basic things in view i.e. Bhangi*-Mukti and environmental sanitation. Thus, the sanitation movement started by Gandhiji and his disciples was bi-dimensional or tri-dimensional. Bi-dimensional in the sense that it aimed at liberating scavengers through the Bhangi*-Mukti programme and also to check environmental pollution for improving the health and hygiene of the rural folk. It was tri-dimensional in the sense that in addition to achieving the above two objectives, it also provided good manure for farms which would increase land fertility and income. The village sanitation movement, launched by Appa Saheb, had been undertaken by the Maharashtra Gandhi Smarak Nidhi (MGSN), which was established in 1950. A number of welfare and social service programmes had also been undertaken by this organisation which included the programme of village uplift and Bhangi*-Mukti scheme. The MGSN is still carrying out these programmes. In 1958, the programme of village sanitation was taken up by this organisation and the social workers or employees who were given the work of scavenging and cleaning latrine were non-scavengers. It was disclosed by the responsible persons in the MGSN that a non-scavenger was given the task of cleaning latrine even in the office and other establishments of the MGSN. It simply means that along with launching the sanitation movement, the Maharashtra Gandhi Smarak Nidhi had also taken up the programme of 'Bhangi*-Mukti'. In this way a sanitation movement had been launched by MGSN in the context of scavenging.

The sanitation movement, initiated by Gandhiji at Calcutta Congress in 1901 and later at Sabarmati Ashram in 1918 by his followers in Gujarat and Maharashtra, created a favourable climate for rural sanitation. The experiments conducted by Appa Saheb Patwardhan in Ratnagiri and the actual plan of action implemented by him proved a pioneering step in the sanitation movement in India. In the following years, a number of experimental works and programmes of action on sanitation were taken up in the country with the assistance of the Governmental machinery and some international agencies. These experiments were landmarks in the history of sanitation movement. The important ones among them had been those that were carried out in Poonamalli (Tamil Nadu), Singur (West Bengal), Bukkshi-ka-Talab, Lucknow (U.P.) and Najafgarh (Delhi).

Several national as well international agencies and organizations have made valuable contributions in respect of research and innovative ideas for developing the technology of pour-flush latrines with leaching pits for onsite disposal of excreta with a view to making them affordable and acceptable by the people concerned, and to promote the cause of sanitation improvement in the country. However, these were primarily concentrated for betterment of rural sanitation and were carried out during the latter part of the nineteen forties and after Independence in the nineteen fifties. It is in the early years of the nineteen seventies that application of low cost latrine was tried out in the urban areas by some non-governmental organizations which were more or less rural in character. However, the real breakthrough in the adoption of low cost pour-flush latrines suited to the congested urban areas was achieved by Sulabh International in the mid seventies. A number of such agencies and organizations have also made valuable contribution to the liberation of scavengers and their rehabilitation. The Central and the State Governments have undoubtedly played a very significant role in these activities.

The sanitation situation in rural India as mentioned in the annual reports of Sanitary Commission to the Government of India indicates that in the early part of the century, trench latrines were experimented and used in the Tea Estates. Some rich villagers had put up latrines of one kind or another but none of the prevalent types were satisfactory for extensive adoption. No serious attempt was made to solve the excreta disposal problem in rural areas till 1935.

With the passing of the Government of India Act of 1935, Health Administration in rural India became the responsibility of District Health Officers. But the task of promoting latrines was left to the discretion of Sanitary Inspectors who had multifarious responsibilities. With the result, excreta disposal in villages received very little attention. Sporadic efforts were made in developing latrines in different regions, the Gopuri and Nasik type in Bombay state, the Hagevu in Mysore, and the borehole in plains of U.P., Bihar, Punjab, West Bengal and Delhi. The Wardha type of latrine consisting of shallow trench was advocated by Mahatma Gandhi. He advocated removing and covering the excreta with earth after defecation by the individuals themselves, no matter to which caste they belonged. It served as an important example of an improvised method for the prevention of unsanitary conditions.

Between the years 1940-1945, increased attention was devoted to the improvement of borehole latrines in the rural west Bengal through the Rural Health Center, Singur attached to the All India Institute of Hygiene and Public Health, (AIIH&PH), Kolkata. In the late forties the borehole latrines were modified into dug well latrines with a water seal trap developed by Prof. K. Subramanian, Head of Sanitary Engineering Department of the Institute. This was the first introduction of pour flush latrines. The provision of a trap with waterseal and casting squatting plates was developed at Singur Health Centre. Systematic and intensive investigations were also carried out by AIIH&PH on the extent of travel of pollution from such dug well latrines with a grant provided by the Rockefeller Foundation.

In the post independence period there was increased emphasis on research activities in regard to the promotion of rural latrines. The Planning, Research and Action Institute (PRAI), Lucknow undertook research with the assistance of W.H.O. and the Government of U.P. as part of their work for improving environmental sanitation. The Institute introduced the concept of using double pits. Another important development was the establishment of the Research-cum-Action (RCA) Projects in 1956 by the Union Ministry of Health with the assistance from the Ford Foundation. The Indian Council of Medical Research also sponsored research units in Kerala, West Bengal, Bombay and Punjab to study the types of latrines suitable for different regions. Some useful work was also done during this period by American Mission in Barapalli in Orissa and by the Kerala Government under Norwegian and World Health Organisation projects. The States of Gujarat and Punjab undertook projects of the construction of latrine for Hookworm Control. Organisations like American Friends Service, Ludhiana (Punjab), WHO project at Ranagram (Mysore), Sanitation Wing at Gandhigram (Madras), the National Environmental Engineering Research Institute (NEERI), Nagpur, the Central Building Research Institute (CBRI), Roorkee, WHO, UNICEF, UNDP, World Bank and several bilateral agencies have contributed to the development and promotion of sanitation programme in the rural areas of the country.

DEVELOPMENT OF LOW COST LATRINES

The research and development carried out by different organisations at different times have already been stated in the foregoing paragraphs. In course of the above R&D works, the following type of latrines were developed.

BORE HOLE LATRINE

This type of latrine was developed by the All India Institute of Hygiene and Public Health (AIIH&PH) at

Calcutta with the assistance of the Rockefeller Foundation. This latrine was popular during the pre-independence period with Health Units established in the country by the Rockefeller Foundation. The latrine consists of a bore hole of 12" to 16" diameter dug in fairly soft soil by special auger to a depth of about 2 ft. or 3 ft below the ground water table. A concrete squatting plate with a central opening and foot rests is placed over the hole and a suitable enclosure built around it for privacy. The hole is lined with bamboo matting or earthen rings to prevent the caving of soil. The excreta and ablution water in the latrine fall into the hole and undergo anaerobic digestion. With usage the hole gradually gets filled up with digested sludge. When sludge is filled up to 3 ft from ground level, the squatting plate is removed and the hole sealed with earth. A new hole is dug and the squatting plate is removed and a new enclosure is provided for the latrine cubicle.

DUG WELL LATRINE

Bore hole latrines were not very popular because of the limited life and difficulty in constructing them in loose soil and high water table situations.

AIIH&PH carried out extensive studies and developed dug well latrines. The dug well latrine consists of a circular pit of 30" dia. dug up to a depth of 6' – 10' depending on the soil conditions. In the sandy soil, the depth is reduced to 6 ft to 7 ft. The pit is lined with 9" – 12" depth pottery rings made locally and as many rings as necessary are used. A brick or concrete ring is provided on the top of the pit and the squatting plate with a hole placed on top of the pit, is at least 9" above the ground level to prevent the rain water getting into the pit. A suitable superstructure for privacy is provided. When the pit is full, a new pit is made. The squatting plate and superstructure are removed to the new dugwell.

HAGEVU FLUSH LATRINE

The Hagevu Latrine in Mysore is a modified pit latrine. A pit 1 ft 9 inches in diameter is first dug to a depth of 3 ft. After this depth, the diameter of the pit is gradually increased to 2 ft – 9 inches at a depth of 7 to 8 feet and 4 ft at a depth of 16 to 17 ft. If the soil does not permit digging in a slanting direction the pit may be 2.5 feet to 3 ft square or round throughout the whole depth.

GOPURI LATRINE

This latrine was developed in Maharashtra State. The latrine is similar to a pit latrine but the pit is constructed above the ground to facilitate removal of excreta and covering of material from the rear of the pits. Each latrine is provided with two pits which are used alternately. The squatting plate with hole is mounted on one pit while the other pit is not in use, and is covered with galvanised sheet mounted on a frame. A vent pipe is provided for the exit of foul gases. Materials like dry earth, ash, paddy husk or hay are used for covering the night soil collected in the pits. The manure obtained after composting of nightsoil is found to contain about 2 per cent nitrogen.

KHATGHAR TYPE LATRINE

This type of latrine was used in some parts of Maharashtra State. The latrine consists of a masonry tank 4' x 3' x 3' for 10 to 20 users and an auxiliary tank 1'x 1' x 3' to increase the detention period of the effluent. The seat consists of a cement pan at the mouth of which is attached a G. I. plate with a hinge. When the excreta presses the plate it is released into the pit.

BARAPALLI TYPE LATRINE

In an effort to improve environmental sanitation in villages, a pit latrine with water seal squatting plate was developed at Barapalli village, Orissa State. The dimensions of the latrine pit are 30 inches diameter and 3' to 5' deep.

AQUA PRIVY

The aqua privy is a simplified septic tank and has been used in different regions of the country particularly in wayside stations in railways. The privy consists of a tank filled with water into which dips a chute or drop pipe hanging from the latrine floor. The faeces and urine fall through the drop pipe into the tank where they undergo anaerobic digestion as in a septic tank. No more additional water than what is used for anal cleaning enters the tank. There is hardly any overflow of effluent. However a soak pit is provided to trap the same if there is any overflow.

POUR FLUSH LATRINES (SINGLE PIT)

With further research and investigation, the AIIH&PH developed the pourflush latrine with leaching pit (with walls having honeycomb) in the late forties. A trap with waterseal is fixed to the outlet of the pan of the squatting plate placed on the top of the dug well. The waterseal is only 3/4" and the excreta can be flushed with about 2 liters of water. No vent pipe is necessary. The waterseal prevents the odour nuisance of dugwell latrines. A single pit is provided. When the pit gets filled up, a second pit is dug as in the case of dugwell. The squatting plate with pan and trap is placed on the new pit. The filled up pit is covered with earth and excreta gets digested and formed into soil.

PRAI TYPE LATRINE (TWO LEACH-PITS POUR FLUSH LATRINES)

The Planning Research and Action Institute, Lucknow carried out systematic studies on the different features of latrine pan and trap in the mid-fifties. The concept of using double leaching pits (described earlier) with offset arrangement was first demonstrated by this Institute. A curved pipe was provided to serve the second pit by changing the direction of the pipe, so that the latrine cubicle was not required to be shifted. The second pit was to be dug by the side of the first pit when the latter was filled up.

R.C.A. TYPE LATRINE

The Research-cum-Action Project (RCA) developed suitable engineering designs for rural latrines by studying the available latrine designs in the country. The RCA latrine consists of two leach pits 30 inches in diameter each, and 6 ft to 9 ft deep. The squatting plate is 3 ft square with slope of 6" towards the pan. The length of the pan is 17 inches with a slope of 1:25. The trap is 2 3/4" to 3" in diameter with 3/4" water seal. A Y-type junction connects the two pits.

SULABH SHAUCHALAYA

All the above designs were tried out and meant for rural areas. Sulabh International was the first to develop in 1970 a two leach pits pourflush latrine suitable for thickly populated congested towns and cities. The leaching pits are rectangular, divided into two compartments having a common solid wall. The pan having a sharp slope and a trap with 1/2" to 3/4" waterseal are made of cement, concrete, ceramic or GRP. Large scale construction of Sulabh Shauchalayas has been taken up all over India which has gained acceptance of the people.

EXPERT GROUPS ON HEALTH, SANITATION AND ENVIRONMENTAL HYGIENE

In connection with post war reconstruction plans, the Government of India appointed a committee to be known as the Health Survey and Development Committee in October, 1943 under the chairmanship of Sir Joseph Bhore to conduct a broad survey of the existing position in regard to sanitation and health conditions in British India, and to make recommendations for future developments. The Committee studied the situation and suggested ways for its improvement. The report of the Bhore Committee was ready in 1946. In about two years' time another committee, the Environment Hygiene Committee was constituted by the Government of India, Ministry of Health in June 1948 under the Chairmanship of Dr. B.C. Das Gupta, primarily to consider the steps to be taken to implement the recommendations of the Health Survey and Development Committee in regard to Environmental Hygiene. The main recommendations of the above two Committees as far as excreta disposal is concerned were as follows:

I. Bhore Committee

The manual handling of nightsoil during its collection and disposal should be rendered unnecessary as far as possible by (i) sewering all towns with a population of 50000 and above in ten years at a cost of Rs. 60 crores, (ii) providing septic tanks with soil absorption system in smaller places, urban and rural and (iii) providing borehole latrines in villages with squatting plates of porcelain or concrete as well as reboring service.

II. Environmental Hygiene Committee

(i) Sewage and sewage treatment in all cities over one lakh population and pilgrim centers and predominantly industrial towns at a cost of Rs. 15 crores, (ii) for unsewered towns conversion of service latrines into septic tank latrines with disposal of excreta in hygienic manner with use of effluent in soil or kitchen garden, and (iii) for villages, waterseal latrines, Mysore type latrines (HAGEVU), borehole latrines and pit privies.

Subsequently attempts to improve method of scavenging were confined to making the disposal of human excreta a cleaner operation by providing scavengers with gloves, gum boots, collection implements and covered buckets and wheel barrows.

It was the V.N. Barve Committee (1949-52) of the erstwhile Bombay State which, for the first time, made some detailed recommendations that were circulated to all State Governments for their attention by the Ministry of Health in September, 1955. The Estimates Committee of the

Parliament in its 48th Report in the late fifties asked the Home Ministry to impress upon the State Governments the urgency of eradicating the practice of carrying nightsoil as headload and recommended financial assistance to local bodies for providing handcarts and wheelbarrows etc. for the purpose.

The Ministry of Home Affairs in its letter of October 1956 drew attention of the State Governments to the Report of the Backward Classes Commission and pointed out that the Condition of Bhangis* (Scavengers) was extremely miserable and the betterment of this unfortunate section of our people was one of the crying needs of our country. When compared with the average income of the common people, the earnings of the scavengers attached to municipalities, local bodies, etc. may not perhaps be comparatively very low and yet, the condition of these people is really miserable. Drinking and gambling habits, lack of education and lower standard of living are considered to be mainly responsible for many of their ills. Housing conditions of Bhangis* badly need improvement. This objective could be achieved by dispersing them in batches so as to provide clean residential accommodation for them among other sections of the people and also by giving them comprehensive training in methods of cleanliness and hygiene. This process, which may be spread over a number of years, would, to a great extent, help Bhangis* to be absorbed in society. The stigma of untouchability would only then be effectively removed.

This was followed up by a number of communications during 1956-60 for 50 percent grant-in-aid for purchase of necessary equipment and implements, to the local bodies through the Social Welfare Departments of the Central and State Governments which have continued to handle all matters relating to 'Harijan Welfare'.

A landmark in the process was the appointment of an All India Committee under the Presidentship of Mr. N.R. Malkani, a member of the Rajya Sabha. Following the discussions at a meeting of the Central Advisory Board for Harijan Welfare held in October, 1957 under the Chairmanship of the Late Shri G.B. Pant, the then Home Minister, it was decided to prepare a "Scheme of putting an end to the practice of carrying night soil in baskets and buckets as head loads". The Committee after comprehensive study of the problem recognized that the best system for the disposal of nightsoil through mechanized means was the flush out latrines, but unfortunately introduction of this, throughout the country required a lot of funds, and thus, seemed outside the scope of practicability in the foreseeable future. The question, however, arose of providing alternative methods of disposal of nightsoil which, as an interim measure, comparatively involved not much outlay taking into consideration the magnitude of the problem. The Committee suggested that the mechanical and up-to-date methods of cleaning latrines may also have to be introduced and the inhuman practice of doing this work by hand and carrying nightsoil as headload should be obviated as far as possible. Sewers, drains, etc. should be introduced on a large scale.

The committee on Customary Rights on Scavenging (1965-66) was constituted by the Department of Social Welfare, Government of India. The committee found, that where scavenging was not municipalized, the latrines were cleaned privately and one particular scavenger acquired hereditary right to clean such latrines as against another scavenger by an understanding and agreement. A customary relationship also develops with the householder and the scavenger receives payment in some form or the other.

In 1967, with the commencement of the Gandhi Centenary Year, emphasis was laid on emancipation of scavengers. The Government provided 25% subsidy and 75% loan for converting dry latrines into water flush toilets and connection to sewers. But this programme failed to attract community participation.

Efforts were made by Government of India in 1968 by promoting Special Campaign for conversion of dry latrines into water flush toilets with the object of relieving the scavengers from the demeaning task. With that view in mind, it was emphasized that :

(a) No new building should be allowed to be constructed without water flush toilets.

(b) Where sewers are available, the dry latrines should be compulsorily connected to them after conversion; and

(c) In the unsewered areas septic tanks or leach pits should be provided for converting dry latrines to water flush units.

The prevailing conditions and the national concern in this regard were conveyed in a Report of a Committee of the National Labour Commission to study the working and service conditions of Sweepers and Scavengers, 1968/69. The report said: "The question of working conditions of sweepers and scavengers is not merely a matter of employer/employee relations, but an issue that has figured prominently in the social and political thinking of the country for almost a generation. The question of untouchability, caste differentiation, the stigma and repugnance attached to this occupation, the age long suppression, privations and afflictions, suffered by these communities of sweepers and scavengers, involve the social conscience of the whole nation. The uplift of this class of people requires breaking of traditional attitudes and a long process of education not only of the higher castes, but of sweepers and scavengers themselves, to pull them up from the state of despair to which they appear to have a fatalistic approach". The report further stated: "That a section of humanity should be condemned to such degrading conditions is a slur on our culture and disgrace to our society". The Committee made another important observation : "It is little realised that by promoting more mechanised system of nightsoil disposal and introduction of water flush latrines, we would not only be rescuing this depressed section of society from a wretched existence, but would also be solving to a large extent, the problem of public health and sanitation of our rural and urban habitations".

Accordingly, several attempts were made in the 5th Five Year Plan by the Government of India for converting dry bucket latrines into water flush toilets by providing community septic tanks. Some 30 towns were selected for demonstration purposes, with 100% grant from the Government. The project failed as cost of maintenance was too high and was beyond the capacity of the local authorities to bear.

NATIONAL WATER SUPPLY AND SANITATION COMMITTEE (1960-61)

The Ministry of Health, Government of India constituted a committee in April, 1960 to make an assessment of the requirements of water supply and sanitation, both urban and rural, and suggest ways and means for financing these programmes effectively. Since the National Water Supply and Sanitation Programme had a tenure of about five years, the Government of India felt that it would be useful to make a critical review of the progress made so far and assess the nature of the requirements of future water supply and sanitation. The Minister for Local Administration, Tamil Nadu was the Chairman of the Committee and the members were health physicians and public health engineers. As low cost pour flush system had not been tried in the towns and cities by that time, the use of this system had not featured in this report and it dealt only with the sewerage system. It had observed that investment on sewerage had a low priority and except a few provincial capitals, sewerage was an unattainable luxury. It recommended that a sanitary latrine for each village house should be promoted under a sustained programme so that filth borne diseases might be eliminated from the rural areas as soon as possible. The Committee gave several recommendations in respect of the organisational set up, procedural reforms, procurement of materials, and water pollution control programme.

CONTRIBUTION OF THE GOVERNMENT AGENCIES AND NGOs

All India Institute of Hygiene and Public Health (AIIH&PH), Kolkata

The Institute developed the borehole latrine, one of the earliest types of low cost sanitation latrines providing onsite disposal of human excreta. This was developed with the financial assistance of the Rockefeller Foundation of USA, and introduced in the country in 1940. Subsequently Prof. K. Subramanium of the Institute developed the dugwell latrine and in late forties, the pour flush water seal toilet with a single leach pit was developed by him. Thus the concept of pour flush latrines was first contributed by Prof. Subramanium. The design was, however, meant for rural areas.

Later, Dr. Bhaskaran and Prof. Dyer carried out studies on travel of pollution in different hydrogeological conditions and developed suitable types of latrines for water logged and high water table regions in the country. The results of the pollution studies showed conclusively that (i) the extent of pollution from latrines constructed in soils mixed with clay is very limited, (ii) in soils, as in the Singur experimental area, which was made of sand 0.2 mm. effective size and velocity of ground water flow of 3 feet per day, the bacterial pollution did not travel beyond 10 feet from the latrine, (iii) in all the experiments, it was observed that pollution flow from latrines takes place only during the first few months and when the latrine has been in use for sometime the interstices of the soil get filled up, resulting in regression of pollution.

Further studies in regard to improvement in the design of latrine, the geometry of the pit and the life of the pit were carried out by Prof. K. Subba Rao. He established the criteria for the increased volume of the pits needed in high subsoil conditions.

AIIH&PH continued its research and investigation in the field of low cost sanitation, specifically in the various problems relating to pollution from the leaching pits of low cost latrines. In association with International Reference Centre for Waste Disposal (IRCWD), Zurich, and subsequently with Bhaba Atomic Research Centre and Sulabh International, Prof. K. J. Nath and Dr. Indira Chakravarty of the Institute have carried out an intensive study on the travel of pollution from leach pits in different hydrogeological conditions with the help of radio isotopes. The investigation was carried out at Chandannagar, West Bengal. The Institute is also involved, and is organising training courses for the personnel of different disciplines engaged in the low cost sanitation (latrine program) with the assistance of Rural Development Department, GOI, ODA of the U.K. Government and the World Bank. In the past 40 years the Institute has trained large number of sanitary engineers and public health physicians from all over the country in low-cost sanitation during their postgraduate studies. It is the first Institute of its kind to have introduced a course on affordable sanitation technology and has carried out pioneering work in every field of low-cost sanitation

Planning, Research and Action Institute, Lucknow

The Institute was established in the fifties to look into the various aspects of rural development, including health. Rural water Supply and Sanitation formed a major part of its health activities. In these fields, assistance was provided by the W.H.O. and the Government of India and the Government of UP. The W.H.O. assigned a sanitary engineer (an international expert) who was assisted by sanitary engineers of the U.P. Government. Various investigations in the fields of technology were carried out viz. cost reduction, users' problems, software needs, latrine construction, maintenance and usage. The Institute developed a mosaic/concrete squatting pan with a concrete trap in two halves which is fixed to the pan at site. Unlike earlier models they are not cast with the cover slab. These are conveniently fixed to the floor and easy to transport. A dome is provided for the pit cover, thus reducing the thickness and requirement of steel compared to a flat slab. The pan and the trap are popularly known as PRAI type ones and used in rural areas. PRAI's significant contribution was the development of the offset latrine pits. This was an improvement on the design developed at AIIH&PH of pour flush latrine with single pit with the latrine cubicle over it. The latrine cubicle in the double pit system is built very near the pit but not over it. When the pit is full it is not necessary to shift the latrine cubicle. A second pit is dug near the first pit after the first pit is full and the latrine connected to the new pit with a curved concrete pipe in two halves joined with mortar at site. The W.H.O. Sanitary Engineer, Mr. James Arbuthnot introduced the above system in the later part of fifties while he was attached to the Institute. The design is meant for rural areas. However, it was observed later that when the first pit got filled up, it could not be cleaned because of raw excreta on the top. Necessary labour to carry out this type of work was not available in the villages. The second pit could not be dug immediately for a variety of reasons, finance, labour, technical guidance etc. People, therefore, took recourse to defecation in the fields. However, by and large, the single pit system with provision of the second one to be dug up in future was a failure.

Public Health Institute, Poonamallee (Chennai)

The Research-cum-Action project on Environmental Sanitation was started in Poonamallee in 1955. The main objective of the project was to evolve practical methods for promoting environmental sanitation with particular reference to rural sanitation. The study primarily aimed at three activities: (a) designing simple type of latrine suitable for rural areas, (b) experimenting with different methods of health education materials by which rural people could be motivated to accept, construct, use and maintain sanitary latrines, (c) suggest ways, particularly organizational and administrative, under which rural sanitation programme could be promoted.

After three years of experimental work the design of a suitable latrine was finalised. The studies were published in the Indian Public Health Journal 1959 by Dr. Pisharoti. Observations were carried out about the rate of filling of the pit in 54 functioning latrines for a period of 14-24 months. The main factors influencing the per capita sludge accumulation were: (i) pit condition-dry or wet: (ii) porosity of soil; and (iii) cleansing materials used. The study concluded that a capacity to hold 1.6 cubic feet of sludge per person per year was adequate. The Institute also carried out programmes in large number of villages and small towns. Its experience contributed immensely in preparing LCS program for the country.

The National Environmental Engineering Research Institute (NEERI)

National Environmental Engineering Research Institute (Formerly CPHERI) located at Nagpur developed and propagated pour flush water seal single pit type latrine for rural areas. The seats were built as part of the pit cover slab. The pits were located below the latrine seats and the enclosure was to be of any light cheap material e.g. bamboo, bamboo matting. After the pit was filled up, the enclosure was required to be shifted over to a new pit and a new seat was required to be installed. The surface quality of the seats was poor as the seats were made of clay. The appearance was dull and it was difficult to keep them clean. Moreover breakage was high in transporting them. Considering these defects, the Institute decided to have cement concrete seats, which were stronger. These could be cast using wooden moulds in different villages. Being heavy these could not be kept hanging over the pits and hence the system was further changed to off-set the pit method. The idea of construction of two pits system emerged to take advantage of costly and permanent superstructure. Initially a simple Y-Junction was thought of for connecting the two pits.

The Central Building Research Institute (CBRI)

This premier Institute situated at Roorkee, Uttarakhand, has carried out engineering studies as well as testing the structural designs of the pour flush leaching pit latrines. It has worked out the least costly superstructure based on most economical design providing satisfaction to the users. The Institute has also contributed in developing pans, traps and flushing cisterns, with new materials suited to this country.

Indian Council of Medical Research

The Indian Council of Medical Research established a Research Unit, in Singur, West Bengal in 1956. It carried out useful studies on performance of different designs of septic tanks. The objective of these studies was to determine (i) efficiencies of the different designs from the point of view of suspended solids, BOD, and pathogenic organisms, (ii) rate of sludge accumulation and method of dislodging and (iii) per capita contribution of sewage solids. The studies revealed that the septic tank effluent is as strong as city sewage and requires further treatment before it can be discharged on land or in a water course. It was observed in these studies that a large number of hookworms and ascaris eggs escaped in the effluents in spite of high percentage removal of eggs in the tank. Discharge of the effluents from a septic tank, without further treatment, in open drains, or watercourses or its use in agriculture, exposes the population to the risk of worm infestation. The results obtained from these studies emphasized the need for a proper disposal of the effluent either through subsurface having tile facing or a soakage pit.

Housing and Urban Development Corporation (HUDCO)

One important official agency that developed in the post independence era and that made valuable contribution to the cause of environmental sanitation is Housing and Urban Development Corporation (HUDCO). Apparently this organisation is concerned with the development of Urban Housing, but its activities and programmes also include eradication of scavenging and environmental sanitation. Established in 1971, HUDCO has now emerged as the principal financing and technical institution in the field of housing and urban development. HUDCO's main thrust is on weaker sections. As far as its activities in the field of housing are concerned, so far it has promoted several million housing units, out of which a substantially large number are for the low income and economically weaker sections. Most such units have thus been built in the rural areas only. Impressed by the encouraging response to low cost sanitation programme all over the country, largely due to the pioneering contributions of Sulabh International in this field, the HUDCO has, in recent years, started promoting the adoption of two leaching pits system for the low cost sanitation. In order to help achieve the targets fixed by the Government of India during the International Decade on Drinking Water Supply and Sanitation (1981-1991), the HUDCO, since 1983, has included in its programme the provision of extending financial support for low cost sanitation scheme to facilitate the construction of hand flush sanitary latrines and also for the conversion of existing dry latrines into hand flush latrines. This new thrust in HUDCO's official mandate aims at the eradication of manual scavenging in various parts of the country.

The HUDCO is convinced that sanitation facilities for disposal of human excreta are either inadequate or totally absent not only in the urban areas, but in rural areas as well. It is mainly due to the fact that the conventional water borne sewerage system is highly capital intensive requiring exorbitant operational cost, and it also requires a large quantity of water for flushing which is already in short supply. The HUDCO is, therefore, now constantly promoting low cost on site sanitation in the country.

The financing terms have recently been liberalized by the HUDCO and financial assistance is now available up to 50% of the project cost at a uniform interest rate of 6% for all beneficiaries irrespective of their income, to be repaid in 12 years. Over and above, .025% rebate will be allowed for prompt and timely repayment of principal and interest etc. in regular installments.

The interest is calculated from the date of the release of the loan by the HUDCO and is payable quarterly on the 31st March, 30th September and 31st December every year. The HUDCO, however, reserves the right to change the rate of interest on the loan amount or a part thereof, depending upon its actual borrowing rate prevailing at the time of release, by giving prior notice to the borrowers of such variation. Any default in the payment of the loan installment on the due dates will not only lead to loss of 0.25 percent rebate, but also additional interest at the rate of 2.5 percent on such overdue payments.

The HUDCCO considers applications for loan received from Housing Boards, Development Authorities, Improvement Trusts, Local Bodies and such other agencies as are authorised by the government to undertake projects and schemes on basic sanitation facilities. The borrowing agencies have to ensure that they are competent to raise the loan constitutionally and legally for taking up sanitation schemes. The loan applications have to be sent along with a resolution of the Board of Directors of Executive Body of the concerned organisation authorising it to borrow the loan.

The sanitation programme for this purpose covers the following types of projects:

1. Conversion of dry latrines into water seal pour flush sanitary latrines (Sulabh Shauchalayas).
2. Construction of new hand flush sanitary latrines in houses where there are no latrines.
3. Construction of community latrines, public bath and urinals in areas where these facilities are not available.

Such schemes submitted by the borrowers should satisfy the following Criteria :

1. It should be technically sound.
2. It should take into account the safe and hygienic systems of excreta disposal with low cost sanitation methods with improved technological development in this field by various institutions.

The proposal for provision of individual latrines or community latrines with disposal arrangements has to conform to the provision of the established norms in this field. The specifications and standards should also conform to the established norms for the same. The thrust of the basic sanitation facilities program should be to adopt low cost sanitation with maximum number of units to be constructed with minimum financial assistance.

The loan for the scheme can be secured by a Government Guarantee of the Scheduled Bank acceptable to the HUDCO. The first installment is released by the HUDCO after the loan agreement is drawn up and other required formalities completed. The second and subsequent installments are released after the advances made previously are utilized to the satisfaction of the HUDCO and the terms and conditions laid down in the loan agreement are complied with. The plans and specifications are approved by the HUDCO before commencement of the work and for any change or revision of the same by the borrowers, prior approval of the HUDCO is necessary. All the projects have to be completed within the time-frame specified in the loan agreement or within the extended period. The repayment period of loans does not exceed 12 years and is be reckoned from the date of drawal of 1st loan installment. Intending borrowers have to prepare a detailed project report taking into account the following information :

1. Details relating to the initiation and the reasons leading to the proposal of the said schemes.
2. Areas specifically identified where such a program is to be taken up in various phases.
3. Existing excreta disposal and drainage system.
4. Sources of existing water supply and its availability.
5. The eligibility and the confirmed demand from the householders willing to avail of the financial assistance, and
6. The system of implementation indicating the method of supervision and control for successful implementation of project.

Harijan Sewak Sangh

It is a non-governmental organisation which has been rendering invaluable service to the Harijan Community of the country for more than half a century. The genesis of this organisation was inspired by the persistent efforts of Mahatma Gandhi to remove untouchability and it was founded on the basis of Gandhian ideology and approach. Gandhiji was more concerned with the liberation of scavengers from their sub-human profession and the improvement in their socio-cultural condition. The work for the eradication of untouchability was being done for quite some time under the inspiration of Gandhiji. But the struggle was not organised and well directed. For this purpose a national organisation was needed. All India Anti-Untouchability League was formed in 1932, which was later on renamed as 'Harijan Sewak Sangh'.

It was realized by the Sangh that the primary requirement for eradication of scavenging would be the conversion of the existing dry latrines into water flush sanitary type which do not require the services of a scavenger to remove the nightsoil from the latrine. The modus operandi of the Harijan Sewak Sangh is to depute its workers not only to motivate the people for adoption of low cost sanitary latrines, but also to supervise and guide the agencies actually engaged in the construction of hand flush latrines and supervise the work at site. For this the Sangh runs a 'Safai Vidyalaya' at the Harijan Ashram, Ahmedabad, for imparting training to the Bhangi Kashta Mukti and Bhangi* Mukti Sewaks in modern and scientific methods of scavenging, construction and maintenance of hand flush water-seal latrines.

Safai Vidyalaya

The Safai Vidyalaya is dedicatingly serving the cause of sanitation. Its activities are primarily confined to the State of Gujarat. The objective of this organisation is bidimensional in the sense that on the one hand, it aims at achieving the target of environmental sanitation, and on the other, it struggles for the liberation, rehabilitation and uplift of the scavengers. The Harijan Sevak Sangh established the Safai Vidyalaya in Ahmedabad in 1963 at Sabarmati Ashram. Padma Shri Ishwar Bhai Patel was appointed as its principal.

The Vidyalaya has carried out social, economic and educational programmes for the scavengers in particular and the weaker section of the society in general. It is, however, active in the state of Gujarat only. The Vidyalaya has involved itself in preparation of mass literature to popularise the sanitation programme and has installed proto type models of various types of latrines within its complex as training aids. Its activity includes training programmes in respect of the methods, approach, use of improved implements for cleaning latrines etc. to all concerned, namely engineers, masons, sweepers and scavengers. Moreover, sanitary inspectors of Primary Health Centers in rural areas and social workers engaged in the programme are also given training. Short-term courses for policy makers and administrators are also arranged at various levels to convince them of the importance of sanitation. This training programme comprises practical demonstration in general cleanliness, technique of conversion of dry latrines into water-borne types, construction of new hand flush sanitary latrines needing no scavenger services, modification of by-law of local bodies and use of improved implement for sanitation etc.

The Vidyalaya has played a vital role in the conversion of dry latrines into the PRAI type. It did not adopt the twin pit but a deep single pit. Demonstrations were held by the Vidyalaya in various Municipalities and Panchayats. The Vidyalaya has helped the Gujarat Government framing bye-laws for the Municipalities and Panchayats and in devising a financial pattern for the successful implementation of the scheme of conversion of dry latrines in Gujarat.

The scavengers trained by the Vidyalaya have been absorbed in other jobs of the Municipalities without creating any problem of unemployment. Though the Vidyalaya has restricted its activities only in the state of Gujarat, it has made a definite contribution to the promotion of sanitation movement in the country.

Gandhi Smarak Nidhi, Pune

The Gandhi Smarak Nidhi is a social organisation which was established in 1950 in Pune. The first Chairman of the organisation was Appasaheb Patwardhan. He was keenly interested in the Bhangi* Mukti Programme and worked intensively for the liberation of scavengers throughout the country. The main areas of the activities of the organisation are village uplift, propagation of Gandhian philosophy, publication of Gandhian literature and the Bhangi* Mukti scheme (emancipation of scavengers).

The Nidhi realized long back that providing better sanitation and proper disposal of night soil in the village was of prime necessity. Hence, a cheaper and hygienic low cost sanitation system had to be evolved. Pioneering experiments in this regard were done by the late Appasaheb Patwardhan in Gopuri Ashram in Ratnagiri District of Maharashtra. Based on his work, the Nidhi decided to propagate the construction of two types of latrines. One is the Sopa type in which the pan which has no waterseal trap but a hinged metal plate is provided at the delivery end before the waste discharges into the leaching pit. It is essentially a leaching pit type hand-flush latrine without a trap. It has a mosaic pan and the disposal of nightsoil is in leaching pits, either twin or single. The other one is 'Naigaon' latrine, which is a modified aqua privy, in which arrangements are made to collect the effluent in containers for use as fertilizer in farms.

Sulabh Movement

The sanitation movement in India entered a new era with the appearance of the Sulabh Shauchalaya Sansthan in 1970 (now known as Sulabh International Social Service Organisation in the country and abroad). The writer founded this organization with the basic objective of Bhangi*-Mukti. He was formerly associated with the Bihar State Gandhi Centenary Celebration Committee which, he joined in 1967. Two years later, in 1969, he was entrusted with the job of Bhangi*-Mukti and looked after the "Bhangi*-Mukti Cell". He was greatly moved by the ideas of Gandhiji on improving the lot of scavengers, He was so much devoted to the cause of Bhangi*-Mukti that he not only served as a social worker in the cell, but developed his own innovative ideas, approaches and plans for the liberation of scavengers. Due to his differences with the Gandhi Centenary Celebration Committee members over the methodology for the implementation of the programme of Bhangi*-Mukti, he resigned from this committee and founded the Sulabh Shauchalaya Sansthan in 1970.

The coming of Sulabh International in the arena of sanitation movement gave a new turn to the programme of environmental sanitation. As is evident from the above facts the sanitation movement was earlier concerned mainly with rural sanitation. The experiments conducted at Poonamalli, Najafgarh, Singur and Lucknow, the programmes implemented on the basis of the findings of the experimental studies carried out in these Institutes, and the social work taken up by the followers of Gandhiji in the country, were all concerned with the rural sanitation programmes. The Sulabh International initially took up the programme of urban sanitation and liberation of scavengers working in cities and towns. Sulabh's action programme was based on the realization that the problem of environmental pollution is very grave and alarming in urban areas due to over-increasing population pressure and the consequent development of slum areas, inadequate facilities for the disposal of garbage heaps and cleaning of drainages, and unplanned building constructions and industrial wastes. The urban people cannot go for open space defecation as people in village do because of non-availability of sufficient isolated open space.

The system of bucket privies for the disposal of night-soil in urban areas, realized Sulabh, is nasty and hazardous to health. Due to bucket privies, the environment remains stinking and the constant contact of flies with night-soil gives birth to diseases like diarrhoea, dysentery, cholera, hook-worm, ring-worm etc. Moreover the system of bucket privies keeps engaged a section of the people, regarded as the lowest in the society, called scavengers, to clean them manually daily. Some arrangements had been made in the past to minimise their insurmountable difficulties and to keep them away from direct touch of the night-soil by providing hand gloves, gum-boots, trollies, etc. but somehow these arrangements could not work. Due to bad condition of the roads, the scavengers find it easier to carry night-soil on the head rather than push the trolly. After three or four months, hand gloves, gum-boots etc. wear out and they continue to clean the night-soil manually. The containers of bucket privies also develop holes and the night-soil gets scattered all around the dwelling houses and creates a problem both for the dwellers and cleaners (Pathak, 1981:9-10).

It is evident from the above remarks made by the writer himself in one of his earlier books – *The Simple Idea that Worked* – that the dry latrine system gives birth not only to the problem of the scavengers who are to undertake a sub-human and demeaning occupation to earn a living. To clean and carry night-soil of one person by another for earning livelihood is highly disgusting and amounts to an act of gross injustice. On the one hand, the practice of scavenging and cleaning dry latrines is a blot on the social system and, on the other, it is a basic cause of environmental pollution. Due to the continued existence of dry service latrines, the intensity of the problem is increasing. Notwithstanding the demerits of the bucket privy system for the disposal of night-soil in the urban areas, in the absence of an alternative, the growing urban population has had to content itself with this system. As a result, the number of persons using bucket latrines has been increasing year after year.

With the phenomenal rise in the population of towns and cities, even the increasing number of service latrines has failed to cope with the rising demand of the urban people. What we find in the urban areas now is that even the bucket privy system is beyond the reach of the urban poor, the pavement dwellers or those living in slum areas. These people still use open fields wherever and whenever lonely places are available for the defecation purpose. The data collected in this regard show that 40 million people or approximately seven million households (1981-82) have no latrine facilities in the country (Pathak, 1982: 6-7). It means that both the service latrine system and open space defecation add to the gravity of environmental pollution. The possible solutions, which were adopted in cities and towns before the appearance of Sulabh Shauchalaya, were sewerage system and the septic tank system.

As regards the sewerage system out of 5,161 urban habitats in the country, only 232 towns were given this facility. Moreover, in most of these 232 towns provided with sewerage system, coverage was partial benefiting 28% of the population. In addition, the sewerage system is very costly and the economic condition of the country does not allow to have a plan of covering all the cities and towns by this system of disposing night-soil.

It has been estimated that if the entire revenue of the Central and State Governments of the next 50-60 years is exclusively utilised for provision of underground sewerage, even half the population of this country would hardly be covered. In the state of Bihar, for instance, the sewerage system was introduced in Patna in 1939. So far a very small portion of the population has been covered. It is obvious that in view of the heavy expenses involved in the construction and maintenance of sewerage system, it cannot be adopted on a large scale. The developing and under-developed countries cannot afford to have this system for the proper hygienic disposal of human waste.

The septic tank system, less costly than sewerage but costlier than the other option already available in India, has found favour only with 14 percent of the urban population. This system was developed way back in 1460 by one Mr. Harrington of England but even after the lapse of 545 years its adoption is very low. In addition to cost factor, septic tank needs more space for construction, which often becomes a problem in a small house in cities due to scarcity of land, especially in congested area. Besides polluting the environment and causing health hazards, the septic tank needs the services of the scavengers to clean the tank when it gets filled up due to the presence of raw night-soil along with the decomposed one. The day the septic tank is cleaned the entire locality remains full of stink, and the cleaning has to be done by the scavengers. Therefore, this system cannot eradicate manual scavenging.

Being the founder of the Sulabh Shauchalaya and, later, of Sulabh International Social Service Organisation, the writer got himself involved in the liberation of scavengers programme in 1967-68 during Gandhi Centenary Celebration period. It is indeed that small involvement which turned into a scavenging liberation movement. As scavenging and sanitation have close relation, the two-dimensional approach adopted by Sulabh International gained momentum with the passing of time, covering not only 26 States and 3 Union Territories of the country but also reaching beyond national boundaries. Sulabh International not only provided an idea of low-cost sanitation system aimed at liberating scavengers, but has also been working for implementation of various sanitation Programmes. Along with achieving success on the front of converting household dry service latrines into low-cost Sulabh Shauchalayas and building new household low-cost sanitary latrines, the organisation has also started the programme of public latrines, building Sulabh Complexes at different places in cities and towns and maintaining them out of the proceeds taken from the users. Sulabh Complexes have been built in towns and cities of almost all States and Union Territories for use by the people who have no household latrines, for the migrant and floating population, the houseless poor, the wanderers and the adrift whose number far exceeds those who have houses with all sanitary facilities. All Sulabh Complexes are working satisfactorily. They have become successful in cross cultural settings and have received appreciation from different quarters. A number of national and international agencies have recognised, approved and appreciated the utilitarian and practical aspects of low-cost sanitary latrines and the efforts of Sulabh International.

As an NGO, Sulabh had been the first to construct two pit pour flush latrines in several towns of Bihar. These latrines demonstrated that in the thickly congested urban areas also the pour flush latrine can be effectively used as solution to the sanitation problem.

To popularize the programme Sulabh International impressed on the Bihar Government to provide to people willing to build household latrine, 50% subsidy and the balance as loan to beneficiaries through local authorities. It took the responsibility of launching awareness campaign and publicity as well as motivating and educating the households. It also got the applications processed, and loan agreements executed and sanctioned on their behalf. All these activities made the program widely accepted. The organisation constructed latrines on behalf of the beneficiaries, keeping them fully satisfied and also being careful to see that there were no complaints from the local authorities. Thus the organisation offered a complete turn key system taking over all the responsibilities on itself. The methodology adopted by Sulabh International was perhaps the most important factor which attracted people to adopt this low cost toilet or "Sulabh Shauchalaya" as it is commonly called. The latrine conversion / construction programme of Sulabh has led to remarkable success in the promotion of sanitation programme in the country.

The other pioneering work of the Sulabh International is the construction and maintenance of "pay and use" type public toilets with bathing facilities. No one earlier could believe that economically weaker sections of the population in the towns and cities, who are used to open air defection, will willingly pay and use the toilet facilities. A detailed account of the contribution of Sulabh International in sanitation movement in India is provided in Chapter VII.

CONTRIBUTION BY INTERNATIONAL AGENCIES

Along with the government and their agencies as well as the non-governmental organizations, the international agencies like WHO, UNICEF, UNDP and Word Bank have also played an important role in the promotion of sanitation movement in the country. These agencies have not only provided financial assistance, but have also given technical assistance for implementing sanitation programme and training personnel.

The World Health Organisation

W.H.O. has been contributing to the development of environmental health in India since its inception. It has been assisting in all aspects of environmental health, particularly in training, research, sanitation education, community participation and their application in the field. The most significant contribution of the W.H.O. in low-cost sanitation system was provided through the WHO sanitary engineers in the late fifties who conceived the idea of the double pit system for the pourflush latrines with leach pits at the PRAI Institute at Lucknow. In 1977, on learning about the success story of Sulabh International, in low cost sanitation at Patna in Bihar, the WHO sent couple of missions to Patna for studying the situation and possibilities of wider application of this system in the whole country. As a first step, it organized a seminar on "Conversion of Bucket Privies into Sanitary Latrines" at Patna in 1978. The seminar was attended by sanitary engineers, health specialists, administrators, planners and decision makers from several states selected by Government of India, various Institutes and Departments of the Government of India, such as Planning Commission, Ministry of Works and Housing (now named Urban Development), Central Board for Prevention and Control of Water Pollution, NEERI, AIIH&PH etc. The august gathering after deliberating for three days on various aspects of this system, in particular technology and pollution, concluded that this system is very well suited to the Indian conditions and would be affordable, both socially and culturally and could be adopted in urban areas. It was after this seminar that governments of India and various state governments got interested in the low cost system. The credit of the application of low cost pour flush latrine system which is found today in this county should be shared by the WHO.

UNICEF

The UNICEF's area of operation in low-cost sanitation has been only in rural areas except in exceptional circumstances. Its assistance, both technical and financial has been in the execution of projects. By providing low cost pour flush latrines in schools, it had taken a major step in educating the school-going children about the benefits of making use of a latrine instead of open air defecation. Community participation and communication support have been the primary features of the projects taken up by the UNICEF. UNICEF had not taken up the conversion of dry latrines, it has only taken up the construction of new latrines where there were none. It is no doubt a measure of improved sanitation but also of avoiding scavenging in future. Although compared to the need, UNICEF's financial contribution might not have been very significant, but it certainly promoted the methodology to be adopted in the field for involving the community both physically and financially for promotion of low cost sanitation programs. Training in various aspects of sanitation programs has also been one of the major contributions of UNICEF.

UNDP – World Bank

The UNDP took up a Global Project for demonstrating the application of Low Cost Water Supply and Sanitation. The World Bank, as its Executing Agency, established a Technology Advisory Group (TAG) in Washington with different disciplines to assist various countries in the execution of the programme. In June 1977, the Government of India with the assistance of UNDP took up the programme. The Technology Advisory Group (TAG) established a Regional Office with Head Quarter at Delhi that assists the various South Asian Countries to take up such projects. The Government of India suggested that Feasibility Study of Low Cost Sanitation should be taken up in the urban towns of 21 states. Sulabh International's success in converting 50,000 dry/bucket latrines into low cost pour flush latrines (Sulabh Shauchalaya) in thickly populated towns of Bihar was the primary factor which encouraged TAG (India) to make in depth study of the system.

The Technology Advisory Group prepared Feasibility Reports on 211 towns, in 21 States and Union Territories. This was done in close collaboration with the central, state and municipal authorities. 211 towns were selected by the concerned State Govts., and union territories after discussion with TAG (India). For preparing the Feasibility Report house-to-house survey was carried out in 15 lakh houses. Information was also collected about the financial status of the local authorities, organizational structure and technical manpower for implementing the project. During the project period, about 14,000 demonstration latrines were constructed in the project towns, which indicated affordability, acceptability and feasibility of pour flush latrine.

The study took care of the serious misgivings of people and authorities about possible pollution from leach pit latrines of ground water and water distribution pipes and, therefore, carried out elaborate investigations. Initially pollution studies on a limited scale were conducted by the TAG (India) with Dr. T.R. Bhaskaran as consultant in Bihar, Gujarat and Tamil Nadu in 1980. The report was published by the TAG (India) in July 1982. It conclusively proved that in Bihar and Gujarat the introduction of pour flush latrines connected to leach pits did not pose any pollution hazards. In Tamil Nadu, however, the studies indicated that there was likelihood of bacteriological pollution traveling upto a distance of 8 meters. Subsequently another study was conducted by TAG (India) in collaboration with International Center for Waste Disposal, Switzerland, and the All India Institute of Hygiene and Public Health (AIIH&PH) at 'Chandan Nagar' near Kolkata. Yet another study was conducted in collaboration with Kerala Pollution Control Board (Kerala) in soil with coarse sand. It stands to reason from all the above studies that in the vast majority of soil conditions prevalent in the country, the pits can be as close as 3 meters from the drinking water source. However, a distance of about 10 meters may be kept between the pit and the water sources as a matter of safety. In areas with very coarse sand and high ground water velocity, an envelop of 0.5 meter fine sand all round the latrine pit and 0.3 meter below may be used to protect the ground water from contamination.

The TAG (India) conducted a study for the structural stability of the design of the pits suggested in the Feasibility Reports at the Central Building Research Institute (Roorkee). It developed the glass fiber squatting pan and HDPE traps after several investigations and trials with manufacturing and discussion with experts and Research Institutes. In a collaborative programme with National Swedish Institute of Building Research (SIBR) and Central Building Research Institute (CBRI) Roorkee, it evolved a modified squatting pan and trap and also developed a flushing cistern of 1.5 litre capacity. Nearly 500 pans, traps and flushing cisterns were tested at the various Research Institutes and by the State Government in the field.

The TAG (India) prepared a manual on Design, Construction and Maintenance of Low Cost Pour Flush Water Seal Latrines in India which is particularly applicable in the urban area. Important areas of investigation carried out for the Feasibility Report of 21 towns were 1) assessment of pollution hazards; 2) evaluation of ongoing latrine conversion program in Bihar, Gujarat and Tamil Nadu; 3) Optimisation of technology of pour flush latrines; 4) Impact of latrine conversion programme on scavengers and their rehabilitation; 5) Evaluation of community latrines in Bihar; 6) Establishment of design criteria of pour flush latrines; and 7) Municipal, institutional, financial and legal study of the project towns.

The TAG (India) prepared the design requirements for pour flush latrines to be constructed in rural areas of India with the assistance of an Expert Committee, constituted by it. The recommendations were published in the Report of the Committee on Design Criteria for Pour flush Water seal Latrines in Rural Communities in India.

Encouraged by the results of the urban study by the TAG, the Government of India requested the UNDP and UNICEF to take up a study on feasibility of large scale rural sanitation programme in the country in collaboration with the World Bank (TAG). Accordingly the TAG (India) carried out the study in 3600 villages selected from 13 states in the country. The study involved:

i) Analysis of the institutional framework of the department in selected states responsible for rural latrine activities and the NGOs with potential for implementation of large scale programme to meet the decade objectives.

ii) Study and Evaluation of earlier Rural Sanitation Programme.

iii) Determination of appropriate type of sanitary latrines in relation to cultural practices, hydrogeological conditions, socio- economic conditions and local resources,

iv) Evolving alternative financial pattern for making latrines affordable giving due emphasis on community participation.

v) Women's participation in rural sanitation.

The Technology Advisory Group (India) organized 4 conferences on low cost pour flush latrines to create awareness among the engineers, scientists, administrators and decision makers. It felt that unless they were convinced about the suitability of this system, it would not be possible for the Government to embark on a country wide programme for conversion of dry/bucket latrines to water flush system. It was necessary to discuss the various options for improving the sanitation and to adopt the system which the Central, State and Local Governments as well as house holds could afford. It had to be acceptable and affordable.

The first conference was an International Seminar on "Low Cost Techniques for Disposal of Human Wastes in Urban communities". It was held in Calcutta in February 1980 in association with the Institution of Engineers (India) on the occasion of its Diamond Jubilee Celebration. It was attended by 33 foreign delegates and over 300 participants connected with the programme from India, representing all the States, the Government of India, Head of local governments, International and Bilateral organisations & NGOs. They comprised engineers, sociologists, scientists, administrators, planners and decision makers. There was sizeable representation from the local authorities, who are primarily responsible for the sanitation of towns and cities. Almost all the international delegates and about 100 from this country went to Patna after the conference to see how Sulabh International has converted thousands of bucket latrines into pour flush toilets with leaching pits at site in the most congested areas of the city and particularly inhabited by the low income groups of the community. The next two were Regional Conferences. The first one was held at Udaipur (Rajasthan) in August 1982 and the other at Udhagamandlam (Ooty, Tamil Nadu) in October 1982. The conference at Udaipur was attended by the officials of the Government Of India from Ministry of Urban Development (then dealt by the Ministry of Home Affairs), Planning Commission, Ministry of Finance and the State Governments from Northern, Eastern and Central India, CBRI, HUDCO, NEERI, UNDP, UNICEF, WHO, TAG (I), and two NGOs viz. Rajasthan Institute of Local self Government, Jaipur and Sulabh International, Patna. In Ooty, the state Government officials were from Southern and Western India, along with others.

The fourth was a National Conference at Delhi held in May, 1984. It was attended by all the Central and State Government officials concerned with the programme along with the various Research and Training Institutes of Central Government connected with low cost pour flush techniques.

International and bilateral agencies and non-governmental organizations which were involved in low cost sanitation programme were invited in the conference.

The greatest impact of the UNDP/World Bank programme was that they were able to mould the opinion of the administrators, planners, decision makers and engineers that the adoption of low cost pour flush latrine was possible in different conditions both in the urban and rural areas of the country. It was accepted as the least expensive method for eradication of scavenging. It is affordable, acceptable and can be made easily available to the public in general. It is the feasibility study of the 21 states which was perhaps responsible for spreading the message all over the country and for the wide acceptance of the low-cost pour flush water seal latrines.

CONTRIBUTION OF BILATERAL AGENCIES

During the VII Plan, assistance in the sanitation sector was provided by the Governments of Denmark, the Netherlands, Britain, Germany, Sweden and the EEC. These donors mostly assisted the rural sanitation programmes. The Government of Norway took interest in the health condition of the fishermen living in the coastal areas in Kerala in the fifties. The Indo-Norwegian project was perhaps one of the earliest bilateral projects in the field of sanitation for providing low cost latrines in this country. It was started in the mid fifties. About 250 single pit latrines with pan and trap but without superstructure were constructed under the above projects to benefit the fishermen. Both the pan and the trap were made of concrete. These concrete rings each 600 mm dia and 750mm high were used for lining the pit. The soil is sandy and ground water table is high in the area. Between two layers of concrete rings, brick layers with holes were provided for leaching action. The majority of latrines constructed are not being used at present. The World Bank has now taken up the construction of latrines in this area as per design suggested by the TAG (India).

The Rockefeller Foundation

The Rockefeller Foundation was perhaps the first International Organisation to assist this country in the field of sanitation. In 1927, it started its activities in India under its world wide campaign against health problems, particularly those related to hookworms. It established seven experimental health units in India during 1931-39, and concentrated on the introduction of borehole latrines in the villages of erstwhile Mysore State. It provided finance for establishing All India Institute of Hygiene and Public Health (AIIH&PH) at Kolkata in 1931 which focused on training of medical graduates in public health fields. In 1939, Dr. John Grant of Rockefeller Foundation joined the AIIH&PH as its Director and he strengthened the public Health Engineering Department of the Institute with a field centre at Singur near Calcutta. This was the beginning of the research and investigation in the field of environmental engineering particularly on low cost solution for water supply and sanitation. It is a unique Institute of its kind in Asia even perhaps today, where health specialists in all branches are trained. The Rockefeller Foundation continued to provide fellowships to Indian nationals for training and education in sanitary engineering in the western counties till the end of fifties.

Ford Foundation

The Ford Foundation has also contributed to the development of sanitation program in India. It has particularly assisted in the RCA projects of the Government of India, in Tamil Nadu, Najafgarh and Singur (Calcutta) and financed experts for some of the projects

Chapter - VI

THE CURRENT SANITATION SCENARIO

The current sanitation situation in the country is reflected in the Census of 2001. The information made available shows that out of the total population of 1027 million (742 million in rural areas and 285 in urban areas), 303 million people (29.5%) have access to adequate sanitation : 219 million (17.4%) in rural areas and 174 million (61.0%) in urban areas. This leaves about 724 million (70.5%) people who have no access to adequate sanitation : 613 million (82.60%) in rural areas, 111 million (39.0%) in urban areas. There are 12,20,78,136 households i.e. 63.06% of the total number of households where no latrine of any kind is available. Out of these, 10,79,67,200 (78.01%) are in rural areas and 1,41,10,936 (26.03%) are in urban areas. The Statewise distribution is shown in the table below :

Sl. No.	State/Union Territory	Total	Percentace to Total Number of Households	Rural	Percentace to Total Number of Households	Urban	Percentace to Total Number of Households
	INDIA	**12,20,78,136**	**63.06**	**10,79,67,200**	**78.01**	**1,41,10,936**	**26.03**
1.	Andhra Pradesh	1,12,90,651	67.00	1,03,75,523	81.85	9,15,128	21.92
2.	Arunachal Pradesh	92,908	43.69	86,630	52.66	6,278	13.04
3.	Assam	17,45,019	35.35	17,06,369	40.43	38,650	5.40
4.	Bihar	1,12,99,239	80.80	1,08,98,416	86.08	4,00,823	30.30
5.	Chhatisgarh	35,59,319	85.79	31,85,084	94.82	3,74,235	47.40

6.	Gujarat	53,42,498	55.39	46,11,438	78.34	7,31,060	19.45
7.	Goa	1,15,494	41.36	72,892	51.78	42,602	30.77
8.	Haryana	19,58,837	55.50	17,50,950	71.33	2,07,887	19.33
9.	Himachal Pradesh	8,25,915	66.57	7,93,318	72.28	32,597	22.78
10.	Jammu & Kashmir	7,27,166	46.86	6,75,923	58.20	51,243	13.12
11.	Jharkhand	39,05,908	80.32	35,52,620	93.43	3,53,288	33.32
12.	Karnataka	63,95,107	62.50	55,13,914	82.60	8,81,193	24.77
13.	Kerala	10,54,438	15.98	9,22,529	18.66	1,31,909	7.98
14.	Madhya Pradesh	83,00,207	76.01	73,98,577	91.06	9,01,630	32.26
15.	Maharashtra	1,23,74,681	64.09	89,91,687	81.79	33,82,994	41.92
16.	Manipur	71,441	17.96	66,692	22.50	4,749	4.68
17.	Meghalaya	2,05,102	48.80	1,97,479	59.90	7,623	8.41
18.	Mizoram	17,681	10.98	16,077	20.25	1,604	1.96
19.	Nagaland	97,731	29.43	93,809	35.35	3,922	5.87
20.	Orissa	66,97,920	85.10	62,59,607	92.28	4,38,313	40.31
21.	Punjab	18,40,723	43.16	16,39,936	59.08	2,00,787	13.48
22.	Rajasthan	66,33,405	71.0	61,11,318	85.39	5,22,087	23.88
23.	Sikkim	38,354	36.61	37,285	40.64	1,069	8.21
24.	Tamil Nadu	91,90,806	64.84	70,86,871	85.64	21,03,935	35.66
25.	Tripura	1,22,819	18.55	1,19,096	22.06	3,723	3.04
26.	Uttar Pradesh	1,76,65,175	68.57	1,66,31,504	80.77	10,33,671	19.99
27.	Uttaranchal	8,69,334	54.80	8,18,161	6.83	51,173	13.11
28.	West Bengal	88,46,138	56.28	81,56,016	73.07	6,90,122	15.15
29.	Andaman & Nicobar	34,138	46.72	28,635	57.67	5,503	23.50
30.	Chandigarh	42,692	21.14	6,704	31.47	35,988	19.93
31.	Dadra & Nagar Haveli	29,655	67.43	27,104	82.67	2,551	22.79
32.	Daman & Diu	19,252	56.05	15,017	67.97	4,235	34.56
33.	Delhi	5,62,940	22.04	62,920	37.11	5,00,020	20.96
34.	Lakshadweep	998	10.80	367	6.85	631	16.22
35.	Pondicherry	1,04,445	50.05	56,732	78.57	47,713	34.96

In respect of the distribution of households by the availability of service latrines, the Census says that there exist 1,32,10,867 service latrines, out of which 62,31,008 are in the rural areas and 69,79,859 in urban areas. The following table shows the statewise distribution :

		SERVICE LATRINES					
Sl. No.	State/Union Territory	Total Number of Service Latrines	Percentace to Total Number of Households	Number of Service Latrines in Rural Areas	Percentace to Total Number of Households	Number of Service Latrines in Urban Areas	Percentace to Total Number of Households
	India	**1,32,10,867**	**6.9**	**62,31,008**	**4.5**	**69,79,859**	**13.0**
1.	Andhra Pradesh	10,55,649	6.32	3,97,842	3.13	6,67,807	16.0
2.	Arunachal Pradesh	41,321	19.43	28,418	17.27	12,903	26.81
3.	Assam	2,37,226	4.80	1,70,485	4.03	66,741	9.33
4.	Bihar	6,76,198	4.83	4,79,431	3.78	1,96,767	14.87
5.	Chattisgarh	1,20,935	2.91	52,901	1.57	68,034	8.61
6.	Gujarat	4,61,190	4.78	1,34,782	2.28	3,26,408	8.68
7.	Goa	27,991	10.02	11,911	8.46	16,080	11.61
8.	Haryana	3,98,838	11.03	1,49,461	6.08	2,49,377	23.19
9.	Himachal Pradesh	91,974	7.41	69,777	6.36	22,197	15.51
10.	Jammu & Kashmir	4,18,046	26.94	2,61,231	22.49	1,56,815	40.16
11.	Jharkhand	2,75,714	5.67	84,517	2.22	1,91,197	18.03
12.	Karnataka	5,61,113	5.48	2,17,245	3.25	3,43,868	9.66
13.	Kerala	4,26,102	6.46	3,24,374	6.56	1,01,728	6.15
14.	Madhya Pradesh	6,09,690	5.58	1,97,601	2.43	4,12,089	14.74
15.	Maharashtra	8,27,338	4.33	2,92,008	5.38	5,35,330	6.63
16.	Manipur	25,829	6.49	17,925	6.05	7,904	7.80
17.	Meghalaya	34,995	8.32	21,490	6.51	13,505	14.91
18.	Mizoram	11,690	7.26	4,337	5.46	7,353	9.01
19.	Nagaland	52,949	15.94	30,410	11.46	22,539	33.78
20.	Orissa	1,68,277	2.13	90,510	1.33	77,767	7.15

21.	Punjab	5,16,632	12.11	2,26,487	8.16	2,90,145	19.47
22.	Rajasthan	6,16,762	6.60	2,38,036	3.32	3,78,726	17.32
23.	Sikkim	5,189	4.95	4,816	5.25	373	2.86
24.	Tamil Nadu	6,56,257	4.63	2,01,730	2.43	4,54,527	7.70
25.	Tripura	50,701	7.65	39,650	7.34	11,051	9.03
26.	Uttar Pradesh	33,89,803	13.75	18,40,724	8.93	15,49,079	29.91
27.	Uttaranchal	1,75,163	11.04	99,685	8.33	75,478	19.34
28.	West Bengal	8,22,328	5.23	5,14,682	4.61	3,07,646	6.75
29.	Andaman	8,326	11.39	5,737	11.55	2,589	11.05
30.	Chandigarh	18,048	8.94	3,098	14.54	14,950	8.28
31.	Dadra &	308	0.70	67	0.20	241	2.15
32.	Daman & Diu	716	2.08	537	2.43	179	1.46
33.	Delhi	4,11,861	16.12	18,586	10.96	3,93,275	16.49
34.	Lakshadweep	582	6.30	83	1.55	499	12.83
35.	Pondicherry	5,126	2.45	434	0.60	4,692	3.43

In terms of access to improved situation, the following table shows the percentage to total number of households :

Sl.No.	State/Union Territory	Total	Rural Areas	Urban Areas
	INDIA	29.5	17.40	60.70
1.	Andhra Pradesh	26.66	15.0	62.06
2.	Arunachal Pradesh	36.86	30.05	60.12
3.	Assam	18.79	55.52	85.25
4.	Bihar	14.35	10.11	54.81
5.	Chhatisgarh	11.28	3.59	43.97
6.	Gujarat	39.82	19.35	71.86
7.	Goa	48.61	39.74	57.61
8.	Haryana	33.20	23.07	57.47
9.	Himachal Pradesh	26.01	21.36	61.72
10.	Jammu & Kashmir	26.19	19.30	46.71
11.	Jharkhand	14.00	4.33	48.63

12.	Karnataka	32.01	14.13	65.54
13.	Kerala	7.7.55	74.77	85.85
14.	Madhya Pradesh	18.40	6.50	52.90
15.	Maharashtra	30.58	15.54	51.43
16.	Manipur	75.53	71.44	87.50
17.	Meghalaya	42.86	33.57	76.66
18.	Mizoram	81.75	74.27	89.02
19.	Nagaland	54.62	53.17	60.33
20.	Orissa	12.75	6.37	52.52
21.	Punjab	44.72	32.75	67.04
22.	Rajasthan	22.39	11.27	58.77
23.	Sikkim	58.42	54.10	88.91
24.	Tamil Nadu	30.52	11.91	56.61
25.	Tripura	73.78	70.58	87.91
26.	Uttar Pradesh	18.2	10.28	50.04
27.	Uttaranchal	34.15	87.79	67.53
28.	West Bengal	38.47	22.31	78.08
29.	Andaman & Nicobar	41.87	30.77	65.43
30.	Chandigarh	69.91	53.98	71.78
31.	Dadra & Nagar Haveli	31.86	17.11	75.04
32.	Daman & Diu	41.85	29.58	63.97
33.	Delhi	61.83	51.91	62.53
34.	Lakshadweep	82.90	91.58	70.94
35.	Pondicherry	47.48	20.82	61.59

In terms of the availability of the type of latrine within house, Census reports that there were 2,20,76,486 pit latrines (11.05%) out of which 1,42,36,297 (10.03%) were in rural households and 78,40,189 (14.06%) in urban households. In respect of water closet, a total number of 3,45,98,446 households (18.0%) had water-borne latrines. Out of these 98,37,054 (7.01%) were in rural households and 2,47,61,392 (46.01%) in urban households.

The Census data reveals that the size and the magnitude of the problem due to lack of safe and hygienic toilets in rural areas is enormous. There are only 22% of the households having toilets, which include about 6 lac dry latrines and the remaining 78% go for open defecation. In respect of the size and the magnitude of the problem in urban areas, the Census figures reveal that toilet facilities are available to 74 per cent of the households which also include about 70 lac bucket privies. The number of households having water closet in urban areas are about 525 million (46.1%) as sewer lines have been provided in 232 towns and that, too, partially. More than 26 per cent of the people in urban areas resort to open air defecation.

Having looked at these Census figures which portray the current sanitation situation in the country, let us look at the policies and programmes of improving the sanitation situation in rural as well as urban areas.

Rural Sanitation

Rural sanitation is important in view of the fact that India is predominantly a country of villages where, according to 2001 Census, 72.2 percent of the total population lives in rural areas while only 27.8 percent lives in urban areas. In terms of absolute numbers, 74.20 crores of people live in rural areas and 28.5 crores in urban areas. The predominance of rural population notwithstanding, the problem of rural sanitation has not been effectively addressed and the amazing part is that there is a dearth of documented data on rural sanitation which can give direction to planning. The available data are neither complete nor clear. However, the status of sanitation in the rural areas has been examined by the Environmental Hygiene Committee (1949), National Sample Surveys and the Census of India 1991 and 2001. The Environmental Hygiene Committee (1949) observed that in villages hardly 5 percent of the houses had latrine at that time.

The National Sample Survey in its various rounds recorded the following percentage of rural people having latrines and no latrines:

Percentage of Households with Access to Sanitation in Rural Areas

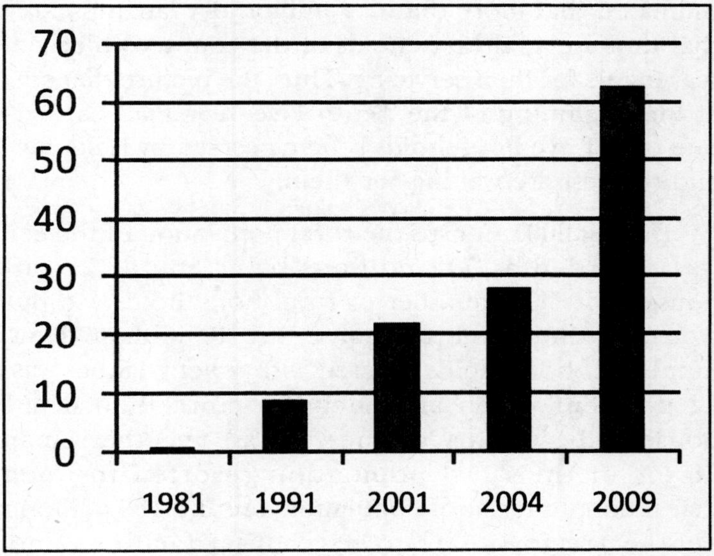

Source: Ministry of Rural Development

The Census of India (1991) had recorded only 2.5% of the population having sanitary latrines. By the end of the Seventh Five Year Plan, only 0.45 million toilets were reported to have been constructed at an investment of Rs. 58.20 crore. During the Eighth Five Year Plan, the programme picked up and 4.34 million toilets were constructed and Rs. 757.62 crore were invested over them. At the beginning of the Ninth Five Year Plan, coverage of rural population with toilet facilities was reported around 16% of the total population. During the Plan period (till March 2001), 4.5 million toilets had been constructed at financial investment of Rupees 1076.54 crore. Achievement for the last several years provides an idea that the average annual coverage of rural population with toilet was about 1% of the total rural population. On this basis it is estimated that at the beginning of the Tenth Five Year Plan, the toilet facilities will have to be made available to about 20% of the rural population. The Planning Commission in 1991 estimated that more than 2.2 million dry latrines existed at that time in rural households of the country which needed scavengers for their servicing. Thus, the picture that emerges at the beginning of the Tenth Five Year Plan, is that only one out of five households is being served by toilet facilities and the rest are waiting for them.

During 2001 Census the rural population of the country was recorded as 742 millions living in 138.27 million households. The number of rural households without any type of latrine within the house was 108 million. Thus, the number of households covered with toilet facilities was only 22 per cent which also includes more than 6 million households with dry latrines (4.5%) and the remaining 78.1% of the rural population resorted to open air defecation. In terms of population, around 160 million rural people are fortunate to have toilet facilities and the remaining, nearing 582 million are still left out.

Low Sanitation coverage in India is primarily due to prevalence of widespread illiteracy and ignorance in rural areas where the people, by and large, are unaware of the elementary principles of hygiene and public health. The people are also not aware of the adverse effects of open air defecation which is responsible for infections and a number of diseases. Such a situation is also due to insufficient motivation/awareness of the people to health hazards and lack of sanitation facilities, or use of dry latrines or open defecation. For people belonging to lower income strata, latrine is still not a felt need and this has resulted in the absence of people's participation. The lack of choice of toilet design, area specific technology, inadequate supporting delivery systems and absence of trained masons, skilled workers and technological manpower are the factors for low sanitation coverage. Additionally by tradition, Indian society and culture value personal hygiene, but give little importance to a clean and healthy community environment. Sanitation is regarded as a matter of individual initiative and not a collective obligation of the community. In this socio-cultural background, environmental sanitation has not received the required priority.

Rural Sanitation Programmes during Successive Five Year Plans

Until the 1970s, there was no special emphasis on rural sanitation. Rural sanitation found only a brief mention in Plan documents. No defined strategy to tackle the problem was evolved because the sanitation programme formed a tiny portion in the health sector. Later sanitation and water supply was transferred to the Ministry of Works and Housing, Government of India and things started looking up. In 1985, Rural Sanitation was transferred to the Department of Rural Development and was launched as part of National Rural Employment Programme, Indira Awas Yojana (IAY) and was also included as Point No. 8 of the 20-Point Programme. It was only in the Seventh Five Year Plan that a centrally sponsored scheme of Rural Sanitation was initiated and was later included in the State sector. The Central Rural Sanitation Programme (CRSP) was launched in October 1986. Since 1987-88, Rural Sanitation formed part of Minimum Needs Programme (MNP) in the State Sector. The programme could not achieve its objective due to poor response from the beneficiaries. Water Supply Schemes received high priority and sanitation could not even make a modest beginning. Without the basic sanitation full benefit of safe drinking water could not be achieved. The programme evaluation studies, however, suggested that by the end of March 1992, the sanitation sector in rural areas improved only marginally: from 0.5% of the households in

1980 to 2.73%. It was estimated that about 22.60 lakh latrines in rural areas had been constructed under various programmes since 1985-86 to 1991-92. However, the 44th round of National Sample Survey (1988-89) revealed that the actual coverage was about 10.96% of the rural households. This made apparent that the balance of 8.23% might have been achieved through private initiative.

Achievements (Latest)	
Household	5.43 crore
Sanitation Complex	16,134
School Toilet	8.57 lakh

In view of the country's dismal situation of rural sanitation coverage (3%), the Planning Commission constituted a Working Group in September 1988 to study in depth the rural sanitation problem for formulating the Eighth Five Year Plan. The working Group was to report on the following aspects of rural sanitation:

1. Technology and Implementation.
2. Community participation, Health Education, Voluntary Organisation etc.
3. Administration and Finance
4. Manpower Development and Training.

Four subgroups were formed for the study with experts in the respective fields. The VIII Plan Approach Working Group Report on Rural Sanitation was presented in April, 1989.

The Report summarised the following important aspects in Rural Sanitation:

1. Nearly 20 to 25% of rural population is keen to install low cost pour flush sanitary latrines on their own provided the correct material and latrine components are made available to them.

2. A minimum of 15% of the allocation should be spent on information, education and communication (IEC) activities.
3. A Central Rural Sanitation Programme (CRSP) with the help of NGOs, voluntary agencies should be able to create a demand from the people.
4. Long term loans should be made available to beneficiaries and motivators. This will create interest in utilisation of the latrines.
5. The project should be on selected district based approach, with intensive activity on education, motivation and delivery system.
6. The programme should be based on individual latrines, and not community latrines, which should be built only in congested areas and should preferably be of 'pay and use' type.
7. Research and Development activities should continue.

Accordingly, the Eighth Five Year Plan guidelines were based on protection of environment, safeguarding of health, promotion integrated approach, including changes in procedures, attitudes, behaviours, the full participation of women at all levels; strengthening of local institutions in implementing and sustaining water and sanitation programmes; the better management of assets; and an extensive use of appropriate technologies.

After having carefully considered the rural sanitation scenario in the country, the Planning Commission had suggested that the sanitation programme be implemented, providing for the following elements :

- All activities under this programme will be undertaken through local bodies / village panchayats and with beneficiary participation.

- Wherever feasible, NGOs will be involved in the implementation of the programme.
- Women will be actively associated with the implementation of the programme.
- Some contribution should be raised from the beneficiaries, at least in the form of physical labour, in order to engender among them the realization that the assets created belong to the local community.

Based on the perception gained from an analysis of the performance of the water supply and sanitation sector in the past and the overall approach and strategies endorsed in the New Delhi Declaration, which was adopted by the UN General Assembly in December 1990, a few important objectives had been visualized for the next five years in respect of sanitation. Some of the these objectives referred to the acceleration of the coverage of rural population with full involvement of the community by the end of Eighth Five Year Plan. It was also felt that Information, Education and Communication (IEC) should be an integral part of rural sanitation programme. In this context, it was thought desirable to adopt a concept of 'total environmental sanitation' and provide guidelines to the rural population in regard to proper environmental sanitation practices, including disposal of refuse, garbage and wastewater through the involvement of local village leaders and community organisations. The Planning Commission had also stressed the need for conversion of all existing dry latrines into low-cost sanitary latrines. It was expected that substantial funds would be mobilized outside Eighth Five Year Plan outlay from the local bodies, the beneficiaries, NGOs and financial institutions like HUDCO, LIC etc.

Rural Sanitation programme could not make much headway despite the recognition of its importance. The initiatives taken during the Seventh and Eighth Five Year Plans were quite insignificant considering the vastness of the problem. The main reasons for the slow progress of the programme were the perception of sanitation as relatively unimportant by the rural people, inadequate financial resources, lack of felt need and lack of people's participation. Although a few successful models of low cost latrines have been evolved by different governmental, non-governmental and international bodies, however, these models are yet to be accepted by most of the rural masses.

There is a pressing need to implement a revitalized programme for rural sanitation, which, as per the guidelines contained in the Tenth Five Year Plan (2002-2007) must have the following elements :

- Preference has to be given to the twin pit model of water-sealed latrines. Alternatively, a single pit model can also be considered for adoption with appropriate changes to suit local conditions.
- School sanitation (providing toilets) should be given highest priority to include safe hygienic habits among school children.
- Village Panchayats should adopt building bye-laws where dry latrines are not permissible. Any latrines to be constructed should be water-sealed type with a leach pit. This will prevent the emergence of the problem of manual scavenging.
- Considering that the programme of installation of low-cost toilets has not made the expected degree of progress, a fresh start is required.
- The idea of setting up State Council for sanitation under urban sanitation sector should also have the mandate for rural sanitation.
- Subsidy for the low-cost household toilets should only be given to below poverty line (BPL) families, and should be on par with subsidy for urban households. For the success of the scheme, a subsidy of 50% of the cost of the unit inclusive of sub and

superstructures for twin-pit pour flush latrines is necessary.

- A quick exercise to arrive at the realistic present costs in different regions, terrains, soil conditions etc, should be carried out so that subsidy amount can be estimated.
- In order to mobilize the required funds for rural sanitation, financial institutions / banks including HUDCO and the National Bank for Agriculture and Rural Development (NABARD) should extend loans at lower interest rates to states for provision of sanitation facilities. Low-cost loan schemes like micro-credit through NGOs should be adequately supported. Various fiscal concessions such as reduced excise duty / sales tax and lower electricity charges should be made available to the manufacturers of low cost sanitary materials.
- Private participation should be encouraged in setting up of building centres and sanitary marts in rural areas to provide cost effective sanitation technology to the rural households.
- The recommendations made with regard to urban low cost sanitation should also apply to the rural segment.
- Creation and maintenance of a record of locally relevant information regarding various technological options, hydro-geological information, availability of building materials, choices in design and implementation etc. at the block level should be organised through the Panchayats, sanitary marts and building centres.
- For the success of the scheme and to overcome the huge problem of insanitary practices in the country, a large programme of education, propagation,

training, designing and development, production and installation needs to taken.

- NGOs should be mobilized to support the programme, especially for supervision, monitoring, training and development work. A suitable provision for the participation of the non-governmental organisations in the sanitation programme should be made in the project costs.

Among the most important elements of rural sanitation package are :

- Safe handling of drinking water
- Disposal of wastewater
- Safe disposal of human excreta. Human excreta is associated with more than 50% diseases
- Safe solid waste disposal
- Home sanitation and food hygiene
- Personal hygiene, particularly, washing one's hand with soap
- Sanitation in community

Approaches that can make the programme a success :

- Decentralization of the programme with full involvement of NGOs and people's participation
- Generation of felt need through motivation and awareness campaign by NGOs, Mahila Mandals and Youth Organisations.
- Adoption of simple and appropriate low cost technology
- Less government subsidy oriented programme can not promote people's participation. It is because of the fact that many of the latrines constructed in rural areas are lying unutilized.

The Existing Central Rural Sanitation Programme (Total Sanitation Campaign) – May 2002

The revised guidelines in January 2004 have made this programme comprehensive with increased emphasis on :

a. Start-up activities
b. IEC (Information, Education and Communication)
c. Rural Sanitary Marts and Production Centres
d. Construction of Individual Household Latrines
e. Community Sanitary Complex
f. School Sanitation and Hygiene Education
g. Anganwadi Toilets
h. Capacity Development Activities to increase awareness and demand generation for sanitary facilities to enhance people's capacity to choose appropriate option through alternative delivery mechanisms.
i. Implementation focused community-led and people centred initiatives.

The strategy is to make the programme Community-Led and People-Centred. The Total Sanitation Campaign has got two new elements. One is that it is demand driven as against supply based which was the pattern earlier. The second is that a campaign approach has been adopted. These two distinguish the existing guidelines from the early guidelines.

The new guidelines aim to cover families below poverty line in the first phase. Subsidy is available to families below poverty line and that is given for simple and less expensive latrine like one leach pit latrine. A duly completed household sanitary latrine comprises only basic low cost unit without any superstructure. The extent of contribution available from Government of India, State Governments and the contribution of the beneficiaries are as below :

Estimated cost of substructure with one leaching pit

	GOI	State	Beneficiary
Upto Rs. 625.00	60%	20%	20%
	i.e. 375.00	125.00	125.00=625
Between Rs. 625.00 and Rs. 1000.00	30%	30%	40%
Above Rs. 1000.00	Nil	Nil	100%

Additional amount can be spent by the beneficiary. The State Governments, Panchayati Raj Institutions, self-help Groups and NGOs can provide additional funds for construction of superstructure and the second pit. The Central subsidy will continue to be admissible with reference to the cost of the Basic Low Cost Unit.

General observations on the Present Guidelines of CRSP (May 2002)

a. In the matter of construction of toilets there is focus on the families below poverty line. Only 27.1% of the population is below poverty line in rural areas. While there is a case for focus on the families below poverty line, it is also a fact that if the Total Sanitation Campaign has to be achieved in full, the guidelines have to be so devised that they can also take care of 70% of the rural population. The basic problem is to generate demand. The Government should find ways and means to generate demand for toilets among all groups of people living in rural areas. It should be examined as to what further steps are required so that there arises a demand for toilets in rural households. It is possible when help is extended either in the form of subsidy, or loans or both to all beneficiaries who want to have toilets in their houses irrespective of income groups. The 'first come first served' should be the criterion for selection of beneficiaries for constructing toilets in their houses.

b. The basic low cost unit-with single leaching pit (without super structure) has been given preference. Without going into the merits and demerits of the single leaching pit or twin pit toilets, it may be mentioned that there should not be a uniform design to be imposed from above. There should be several designs of pour flush toilets meant for the families living below poverty line or those of middle income groups or high income groups. The choice of design should be left to the beneficiaries who will take a decision, keeping in view their requirement and the resources which they are able to mobilize.

c. The existing scheme provides for construction of substructures only. Therefore, the persons below poverty line are not in a position to construct superstructure. The State Government, Panchayati Raj Institutions and other ogranisations are not in a position to provide additional funds for construction of superstructure and the second pit. The result is that the construction of toilet ends with substructures and quite a large number of them are in disuse. As recommended in the Tenth Five Year Plan, the existing scheme should be modified to include superstructure as well.

d. The amount of subsidy should be increased to 50% of the cost of the unit inclusive of sub and super structures for the basic twin-pit pour-flush system as recommended in the Tenth Five Year Plan.

e. It has been experienced that the demand for sanitation among the people below poverty line will be comparatively very low. Their latent demand will have to be stimulated by the non-governmental organisations.

f. The Government should fix up subsidy for hardware preferably on uniform rate and also specify the methodology for transfer of subsidy to the beneficiaries in easy way.

g. Interest free bank loans should be provided to all beneficiaries who want to have toilets in their houses. Thus financial institutions/banks including HUDCO, National Bank of Agriculture and Rural Development should extend interest free loans.

h. In a demand creating approach the consumers are expected to be given more facilities and privileges in the selection of technologies with appropriate price tags. This is intended to be a people's programme and the people should have the real power in the project.

i. While focusing on the families below poverty line, it is also advisable to tap the early adopters available at the village level to start with. The educated, employed and relatively rich may also be contacted simultaneously so that the pace of coverage may increase. If they are provided with latrine, others too will become interested and motivated to have latrines.

j. The experience has been that wherever NGOs have been active the programme has moved faster. The rules/modalities of involving NGOs should be made workable so that they may work and deliver results. For software part the NGOs should be given full responsibilities and allowed to work continuously at least for 5 years.

k. As per guidelines in the Central Rural Sanitation Programme (Total Sanitation Campaign), 25% of cost of the project (5% for start up, 15% for IEC, more than 5% for Alternate Delivery Mechanism, 10% for School Sanitation and 5% for administrative charges on training etc.) are earmarked for carrying out the above vital activities. Besides this, the Programme provides less than 5%, subject to a maximum of Rs. 35 lakhs of the total Government outlay per district for setting up Sanitary Marts and production centres.

As recommended by the National Commission on Population constituted by the Planning Commission, the NGOs selected for implementing the sanitation programme should be allowed 15% of the estimated cost as implementation charges which would include start up and IEC activities including implementation and follow up service. In addition, 10% of the project cost should be allowed for training, support service, and publicity which would include printing of booklets, literature, posters, organizing dramas and nukkad nataks etc. For setting up Sanitary Marts and Production Centres, subsidy as admissible under Government of India Sanitation Programme should also be paid to the NGO.

l. All rules, procedures, monitoring methods should be so designed as to be in conformity with the 'Demand Driven and People centred' strategy of sanitation promotion.

Urban Sanitation

The 54th round of NSS reported that 26% of households reported using no latrine, 35% reported using septic tank, and 22% reported using sewerage system. This indicates that as many as 43% of households in urban areas either had no latrines or no connection to a septic tank or sewerage. As regards waste disposal, 71% of urban households reported removal of household waste by household members, 14% by local authorities, and 12% by private agreement among residents. Forty-seven percent of urban households reported removing of their waste to community dumping spot, and 30%, to individual dumping spots. Ninety percent of urban households reported concern regarding mosquitoes, 66% regarding flies and 50% regarding problems related to foul odour.

i. **Sanitation Coverage in Urban Areas : 63%**

47% Sewers

53% Low-cost sanitation

ii. Urban Areas

1. 30.65 million (35.49%) have inadequate sanitation.
2. 12.04 million (7.87%) do not have access to latrine and defecate in the open.
3. 5.48 million (8.13%) use community latrine.
4. 13.4 million (19.49%) use shared latrine.

Estimates of access to excreta disposal systems in urban areas vary from a low of 48% to a high of 70%. Out of 300 Class-I cities, about 70 have partial sewerage systems and sewage treatment facilities. Levels of sewage treatment are reported to be very low. A study by the Central Pollution Control Board in 1994-95 showed that the total waste-water generated in 300 Class – I cities was about 15,800 million litres a day (mld), while the treatment capacity was hardly 3,750 mld. In the 23 metro cities, over 9000 million litres of sewage is generated daily, of which about 60% is generated in the four mega cities of Mumbai, Kolkata, Delhi, and Chennai. Of the total wastewater generated in the Metro cities, hardly 30% is treated before disposal. Most of the cities have only primary treatment facilities. Thus, the untreated and partially treated municipal wastewater finds its way into water sources such as river, lakes and ground water, causing water pollution. On the other hand, the programme of Low Cost Sanitation, which offers an affordable alternative technology, has made very little progress, despite being linked to the social problem of manual scavenging, which is prohibited by law.

The poor sanitary conditions, particularly in slums are often linked to outbreaks of cholera and gastro-enteritis. Water-borne diseases are a major cause of mortality throughout India and impose a huge burden in terms of loss of lives and productivity. According to a case study, water and sanitation diseases are responsible for 60% morbidity and mortality amongst children.

Low Cost Sanitation

The Centrally-sponsored scheme of Urban Low Cost Sanitation for Liberation of the Scavengers started from 1980-81, initially through the Ministry of Home Affairs and later through the Ministry of Welfare. From 1989-90, it came under the jurisdiction of the Ministry of Urban Development. The main objective of the scheme is to convert the existing dry latrines into low cost pour flush latrines and provide alternative employment to the liberated scavengers. The rehabilitation component is dealt with by the Ministry of Social Justice and Empowerment.

Low cost sanitation is rightly seen as an important solution to the dehumanizing practice of carrying night-soil, and the legislation prohibiting manual scavenging and its enforcement in turn strengthens the movement for installation of sanitary latrines in urban areas. Low cost sanitation is also the appropriate solution where resources do not permit the provision of underground sewerage, or septic tanks. The scheme covers all the households, which have dry latrines and households having no sanitation facilities including households in slums and squatter colonies.

Under the scheme, loan and Central subsidy are extended simultaneously by HUDCO. Subsidy is graded according to economic status, being set at 45% for EWS, 25% for LIG and nil subsidy for middle and high income groups. Loan ranging from 50% to 75% of the cost, is given at 10% interest repayable over 15 years. Loan is also given for construction of community latrines on 'pay and use' principle, shared latrines in slums, housing chawls. Subsidy is limited to the cost of the sub-structure, but the HUDCO loan can be availed for the construction of superstructure also. The loans require State Government guarantee. For management of the programme, HUDCO was required to set up a separate cell, and the regional offices of HUDCO were expected to render assistance to the State Governments in the formulation of proposals. The guidelines provide for the creation of coordination committee at the Central level, as well as coordination committees at the state level. A recommendation has been made to create state-level cells or to nominate an existing institution for channelising the subsidy and HUDCO loans to the implementing agencies. The cell may be registered as a society. In the State of Madhya Pradesh, the Slum Clearance Board was nominated as the nodal agency.

Progress of the Scheme

According to an estimate prepared by a Committee constituted by Planning Commission, there were 400,000 scavengers and 5.4 million dry latrines in urban areas in 1989, and the practice of manual scavenging continues in 3117 towns. Seven states have declared themselves as scavenger free, Goa, Kerala, Gujarat, Manipur, Mizoram, Sikkim and Tripura. Six Union Territories – Andaman & Nicobar Islands, Pudducherry, Chandigarh, Dadra & Nagar Haveli, Daman & Diu, and Lakshadweep – have also declared themselves scavenger free.

The number of households without access to a proper system of removal of human excreta is bound to be much higher than the number of dry latrines estimated by the Committee, because the number of households in urban areas without even dry latrines, who use open places for defecation, is estimated to be 7.3 million. Therefore the number of households in need of low-cost sanitation or community toilet facilities may be as high as 15 million, if not more.

The progress of the scheme as on 15 March 2002 is as follows :

Financial Progress

Schemes sanctioned	- 847
No. of towns covered	- 1,317
Project Cost	- Rs. 1,435.51 crores

Subsidy sanctioned	-	Rs. 486.891 crores
Loan sanctioned	-	Rs. 592.69 crores
Subsidy released	-	Rs. 250.67 crores
Loan amount released	-	Rs. 309.18 crores
Physical Progress		
Total Units sanctioned	-	35,53,585
• Conversion	-	17,05,701
• Construction	-	18,47,884
• Community toilets	-	3,966
No. of units completed	-	14,58,274
No. of community toilets completed	-	2,982
No. of schemes in progress		
• Conversion	-	1,05,619
• Construction	-	2,12,987
• Community toilets	-	185
Towns declared scavenger free	-	387
Scavengers liberated	-	37,430

The data indicate that :

- In the 9th Plan, no loan or subsidy was sanctioned in Assam, Bihar, Haryana, Jammu & Kashmir, Kerala (declared to be free from manual scavenging), Karnataka, Orissa, Punjab, Jharkhand, Tripura, and Andaman & Nicobar Islands.
- Gujarat, Nagaland, Mizoram, Sikkim, Arunachal Pradesh and Himachal Pradesh have also not drawn any loan or subsidy under the scheme since inception.
- Nearly 37 per cent of the loan and 24 percent of the subsidy has been drawn by Andhra Pradesh, which has been consistent both before and during the Ninth Plan.

- Other major beneficiaries have been Haryana, Karnataka, Maharashtra, Madhya Pradesh, Punjab, Rajasthan, Tamil Nadu, Uttar Pradesh, and West Bengal. These ten States account for 92 per cent of both the loan and the subsidy released under the scheme. However, in respect of Haryana and Karnataka, the performance was only in the period prior to the 9th Plan.
- Hardly 13 per cent of dry latrines existing at the beginning of the Eighth Plan have been converted into sanitary toilets during the Eighth Plan period and the first three years of Ninth Plan.

Review of the Programme

Apart from the backlog in numbers of units, there are important issues of user acceptance of the units installed. Evaluation of the scheme was made in 1990 in Maharashtra and Andhra Pradesh, in West Bengal in 1991, and in Uttar Pradesh, Rajasthan, Haryana, Punjab, and Maharashtra in 1994. A 'Benchmark Survey Report for Evolving Community and Women's Participation' for Uttar Pradesh was prepared with the sponsorship of the World Health Organisation in 1994. These evaluations revealed some of the problems of the programme, and judging by the lack of any significant progress, they do not appear to have been adequately attended to in the Ninth Plan period, during which there was no evaluation of the scheme.

The reasons put forth for the slow progress are that the subsidy is limited to the sub-structure; the loan amount is small, and numbers of beneficiaries involved are large, which has made it a difficult task for the local bodies, who generally do not have a high performance in collection efficiency, to recover and re-pay. The states have been unwilling to give guarantees for this reason. On the other hand, in keeping with its policy, HUDCO is unable to provide loans because the municipal finances are in poor shape and most ULBs cannot be funded without state guarantee.

Project management has been weak, with hardly any state establishing an organization capable of taking up the task of propagating the programme and supervising its implementation. Consequently there has been little effort at developing locally relevant options in sub- and super-structure which is responsive to user needs and preferences, using suitable and cost-effective locally available materials, training of local craftsmen to undertake construction of the units, induction of NGOs, quality control, user education, etc. Considerable delay due to bureaucratic reasons has been noticed in the States passing on the project funds to the implementing agencies. Thanks to weak supervision, defalcation of funds has also been reported in some cases.

The Importance of Low Cost Sanitation

Towns which do have sewerage do not often have sewage treatment plants, with the result that water sources are getting polluted. It has been assessed that 80 per cent of pollution is caused by sewage alone. It is infeasible to provide underground sewerage or septic tank latrines in all cities and for all residents. In the first place, highly urbanised, industrialised and densely-populated urban centres may be provided with sewerage, with priority being given to installing sewage treatment plants to prevent pollution of water sources. For the majority of the urban centres, low cost sanitation is the appropriate technology.

Low cost sanitation is not a programme solely for the urban poor or slum population. It has to be propagated as the appropriate solution wherever the costly option of underground drainage is not feasible. In this sense, there is need to offer more options to households that desire sanitation facilities which, while being based on the 'twin-pit pour-flush' model, is in keeping with their needs and capacity to invest. Low cost sanitation is best propagated as a part and parcel of the maintenance of environmental health. This includes the areas of water supply, protection of the environment and preservation of environmental cleanliness, and promotion of health among infants, expectant and nursing mothers, and children. Within a town or city, the proper approach would be to take up a coordinated programme covering sanitation in schools, individual households, and public places with special emphasis on the sanitation needs of the urban poor and slum-dwellers and pavement dwellers. The practice of focusing on water supply to the exclusion of sanitation and waste water treatment, should be given up in the Tenth Plan.

Programme Improvement Measures

Rejuvenating the Implementation Organization : States should set up a State Sanitation Council to be in overall charge of supervision of the programme. The Council should be headed by the Chief Minister or his nominee, and consist of experts, NGOs, representatives of PRIs and ULBs, HUDCO, other institutions in the field of environmental sanitation, health workers, and other concerned officials. The mandate of the Council shall be to carry out extensive propagation of the principles of sound practices in hygiene and sanitation, and supervise the programmes of individual and community toilets, school toilets, pay and use toilets, and water quality. The areas of solid waste management and waste water treatment including drainage, will also be within the ambit of the Council. The Council will implement measures to promote the concept of sanitary latrines while bringing home the indignity of open defecation as well as manual scavenging.

Sanitary latrines have to be promoted as part of an overall health, sanitation and environmental awareness programme, especially among the young. Adequate provision should be made to meet the cost of information, education, communication (IEC) activities under the programme, which seems to be lacking at present. The Task Force had recommended that upto 10 per cent of the budget allocation be earmarked for this component. Out of this, 5 per cent could be utilised for project management and NGO involvement. The cost of training of skilled workers should also be met out of this provision and training should be organised locally with the help of NGOs, Building Centres, and technical institutions. A part of the provision should be used for organising safai karmacharis (manual scavenging workers) to undertake construction of LCS units.

Use of sub-standard materials and construction by untrained workers, in order to cut costs, may result in units which do not fulfill the user expectations, and these may be discarded. There is a minimum water requirement without which the unit cannot be kept clean and may lead to bad odour. Similarly, the space requirements, and suitability of soil must be taken into account before starting work on installing the unit. The guidelines prescribe the distance to be maintained from a potable water source to avoid contamination, and this must be fulfilled. The users need to be trained to keep the units clean and free from odour, and regarding the resting period and removal of the contents of the pit. The superstructure must cater to the comfort of the users, for protection against rain, privacy, etc., and has to meet the minimum criteria of acceptability. These issues will have to be tackled to the satisfaction of the users before the programme can make headway. Strict standards for the construction of the LCS units need to be prescribed and followed in implementation.

The subsidy programme should, however, be based on the cost of the basic twin-pit pour flush model for a small household. Subsidy should be extended to cover the superstructure and substructure. The unit cost should be determined on the basis of local costs, and the loans and subsidy provided accordingly. The householder should have the option to design a unit that meets his requirement. It is suggested that the financing pattern of 50 per cent subsidy and 50 per cent loan, which existed prior to 1988-89 should be restored during the Tenth Plan with the further modification that it should be available to both the sub-structure and the super-structure.

There should be no distinction between the states that have scavengers and those that do not have, because the programme is intended to prevent both manual scavenging as well as open defecation.

HUDCO should establish state-wise units for the management of the programme to provide technical and organisational back-up. The ULBs should be strengthened to undertake the recovery of loan installments.

Community / Pay-and-use Toilets

The scheme of low cost sanitation also includes community toilets. HUDCO has reported the number of units sanctioned, completed, and in progress, as follows :

	Sanctioned	Completed	In Progress
Andhra Pradesh	158	40	58
Karnataka	117	0	0
Maharashtra	2,809	2,663	120
Orissa	10	10	0
Tamil Nadu	372	269	15
Uttar Pradesh	100	0	0
West Bengal	400	0	0

Community Toilet Complex Under Valmiki Ambedkar Awas Yojana

HUDCO is also involved in the implementation of the pay-and-use toilets programme under the Night-shelters scheme, and has reported sanctioning 69 schemes for pay and use toilets, under which a subsidy of Rs 14,000 per seat is provided. This scheme has now been merged with the new scheme of Valmiki Ambedkar Awas Yojana (VAMBAY), as part of the sanitation component of the scheme for which 20 percent of the funds are earmarked. The stand-alone toilet complexes, as distinct from the units provided in slum housing colonies, will continue to be financed under the scheme as earlier.

Community toilets are needed for slum and pavement dwellers, rickshaw pullers and the floating population. However, the experience of maintenance and upkeep of these units by municipal authorities has been dismal, with the conditions turning so insanitary that people prefer open air defecation rather than use the public toilets. The maintenance of the units should, therefore, be handed over to an NGO or a community-based organisation. The beneficiary community should pay a small fixed monthly amount towards upkeep of the toilet and the money should be used for cleaning, water supply, and remuneration of a caretaker.

The construction and maintenance of pay-and-use toilets for the floating population or places such as bus-stands, markets, parks, and other places where people congregate, should be done by an NGO like Sulabh International. In some instances, such installations may even generate a surplus. Despite the possibility of commercial viability of some of the installations, it is premature, in the view of the Task Force, to base the programme on assumptions of viability. In the Tenth Plan, therefore, the construction of community toilets for the urban poor, and of pay-and-use toilets for the floating population should continue to be supported through subsidy from the Central Government.

Towards Achieving the Goal of 'Toilets for All'

The Millennium Development goal of providing adequate sanitation facilities and toilets to all by 2015 can be achieved only when toilet is viewed as a tool of social change. The situation of toiletlessness is the issue that has to be effectively addressed. The increased accent of the campaign for building household and community toilets, and the documentations and discourses, therefore, clearly suggest that the country has to appreciate the civilising impact of toilet, its role in enhancing the quality of life and its centrality in abolishing the manual handling of human excreta. History suggests that during the evolution of toilet system from 2000 BC to date, many notable changes have been brought about by this supposedly innocuous yet realistically important invention. We know fully well how the ordinary inventions in toilet design and technology have stirred history, evoked passions, inspired imagination and kindled hope and vision of toilet providers throughout the world. The invention of toilet has indeed been a 'winning idea' to ameliorate human conditions. The 'winning idea' of toilet invention, evolution, propagation and action-orientation has had demonstrated the power and potential of impacting on the course of human history. While discussing toilet, one ends up discussing virtually everything – culture, economics, science, technology, and even history and geography. The sanitation related writings reveal that toilet is a thing of beauty and not a hideous object.

A deeper study of toilet induced social change reveals that many new things have happened once the world realised that toilet has not merely been a necessary domestic facility, and a source of comfort for good life, but also has resulted in a multitude of critically significant social changes. Some of these are as under :

- Change in socio-cultural attitudes, beliefs, values and practices related to toilet and toilet use.
- Paradigm shift in the perception of toilet – from profanity to prestige.
- Emergence of public sanitation and increased professionalism in sanitation engineering.
- Mounting of public motivation, education and awareness campaigns for promoting the construction of simple low-cost sanitary household toilets

- Improvements in health, hygiene and sanitation status of the developing / low income countries.
- Emergence and growth of pay and use public toilets.
- Development of appropriate low-cost and eco-friendly toilet technology usable on sustainable basis in households and community centres to ensure cleanliness, health and hygiene.
- Integration of sanitation component to slum improvement and upgradation policies and programmes.
- Enabling (through servicing) the weaker sections of society who do not have means, motivation or both to build simple sanitary toilets at their dwellings.
- Toilet provisioning assuming a centre-stage in the process, planning and policy-making in respect of human development infrastructure building programmes and activities to meet the Millennium Development Goals for the fulfillment of which the world stands concerned and committed.
- Development of technology to harness energy from non-conventional sources like human excreta through human excreta based biogas plants.
- Creation of public demand for basic sanitation services, putting pressure on politicians and Government agencies to upgrade and expand the limited outreach of existing sanitation sector.
- Emergence and expansion of the 'toilet market' with accent on manufacturing and marketing simple and fancy toilets, the sanitary hardware and items of toiletry – cleaning materials and tools etc.
- Emergence and consolidation of NGO/civil society involvement in the soft sectors of sanitation development – motivation, demand creation, information, awareness, education.

- Increase in the involvement of socially committed corporate organizations and their help in creating better civic and social amenities.
- Building of comfortable modern public toilets as cornerstones of a better quality of life, comfort, safety and cleanliness - all leading to the uplift of the urban civic life and improvements in personal and public hygiene.

Some of the most significant social changes that have not been adequately amplified are being briefly highlighted here.

A. Toilets have gained Visibility and Space in Media and Public Domain

With the change in the public perceptions and attitudes about toilets and all matters connected with public and household toilets all over the world, toilets have gained visibility and space in media and public domain. Today, national and international conferences, seminars and symposia and high-level international forum meetings including those convened by WHO, UNICEF, UNDP, HABITAT, UNCHS, Water and Sanitation Collaboration Council and International Drinking Water Supply and Sanitation Decade discuss sanitation and toilet centered matters and issues. The World Toilet Association and World Toilet Exhibitions and toilet design competitions are indeed the products of this perceptible change. Had the toilet not got the respectability, visibility and space in media and public domain, we may not have been here to discuss the toilets of the present and future. Another significant symptom of the visibility of toilets in the public domain is that talking about toilet in public gatherings or even on occasions like at breakfast, lunch or dinner table is no longer a taboo. People talk freely, interestingly and absorbingly about matters pertaining to toilet without any sense of hesitation or embarrassment. This, in fact, is big social change worth taking due note of by policy planners and programme executors in the national and international sanitation sectors.

B. Toilets have Increased Involvement of NGOs/Civil Society Organisations

Toilet revolution has also engaged attention and ensured involvement of a large number of NGOs and civil society actors in almost all countries, especially in the developing world. Now they are the important agents for promoting the adoption of innovative approaches and the provisioning of services to support sustainability of sanitation services. They are acting as catalysts of social change in the sanitation sector and are able to bring about a harmonious interaction between technology and social engineering – both being necessary for sanitation to become a social movement. They have encouraged people to realize their responsibilities in relation to sanitation and have made them capable of taking action on their own.

C. Toilets have Promoted Professionalism in Toilet Industry and Toilet Technology.

Toilet inventions, 1890 onwards, have encouraged intense professionalism in companies manufacturing sanitary hardware. Competing with each other in marketing their products, companies come up with colourful bright catalogues on toilets and baths. Evolution in toilet technology and designs makes clear that more and more professional people from business and industry are getting involved in the toilet business. It is no longer a mundane job of ill trained masons, builders and sanitary engineers. On the contrary, toilet business has become a high profile field of entrepreneurship for professionally enterprising architects, civil engineers, and sanitarians of varied expertise. In India, for example, sanitation and toilet related business/ entrepreneurship was a taboo for high caste persons. Until a few decades ago it was a low status business or activity which no respectable person would like to enter. Now things have changed and there is greater professionalism in toilet industry/business and in toilet designing and technology.

D. Toilets have Provided Dignity to Women

The general absence of public and household toilets in most parts of the world, especially in developing countries, including India, put women at great disadvantage. Women resorting to open defecation to relieve themselves did so at considerable cost, embarrassment, fear and insecurity. While men are free to ease themselves in any place they choose randomly, women cannot do that for obvious reasons. They have to have patience, observe self control, draw on their reserve biological and psychological energy, before they are able to answer the call of nature. Women, unlike men are not supposed to do something private in public. And since there are no safe public spaces available in villages, cities and towns, women are supposed to observe and practice self denial. Either way they are condemned. With toilets being built at homes and at public places, women can ease themselves easily. The point I am labouring to emphasize is that it is the invention of toilet that has provided many women a sense of dignity.

Toilets have Served as a Solution to a Difficult Social Problem in India

In India, the innocuous facility of sanitary toilets has served as an effective instrument for the solution of an ancient social problem of the practice of untouchability, especially against a certain section of society who were traditionally ordained to clean human excreta from pit latrines / dry privies. These persons – called by different names, all of them being derogatory and dehumanizing, were subjected for centuries to myriad of human indignities, discriminations, derogations, exploitation and abuse. Mahatma Gandhi, the '*Father of the Nation*', in his life time, struggled hard to heal the woes of these scavengers (toilet cleaners) because his solutions to combating the social evil were based on social moralizing asking people to clean their own latrines, self service to sanitation related jobs, covering of human excreta with dirt/dust, and treating scavengers as one's brethren without trace of any social discrimination. Gandhiji's sermon of moralizing did not succeed because it was not accompanied by a viable technological solution to combat the problem of manual scavenging. The plethora of legislative measures to eradicate the practice of untouchability enacted after independence, following the prohibition of untouchability by the Constitution, could also not make much headway. The legal ban on scavenging also remained ineffective, because of the existence of dry privies serviced manually by scavengers, who had to do the job for want of better occupational alternatives available to them. Sulabh's commitment to resolving this social problem is based on a two fold approach : (1) introduction, promotion and popularization of manual scavenging free technology popularly known as Sulabh Sanitation technology of twin pit pour flush toilets, requiring no manual handling of human excreta by scavengers; and (2) mounting of massive efforts towards the social upgradation and social mainstreaming of scavengers (men and women) who have been freed from their demeaning and socially defining hereditary occupation. Sulabh has so far liberated and rehabilitated 1,20,000 scavengers, provided occupational / vocational training to 7,000 wards of liberated scavengers and generated employment opportunities for such liberated scavengers totalling to approximately 10 million man-days.

Toilets and Toilet-Induced Changes in India

The history of toilets in India is of ancient vintage, much older than the evolution of toilets else where in the world. Vishnu Puran, ancient Indian text, lays down elaborate rules for post defecation ritual. The excavations at the Harrappan settlement at Nausharo (2000 BC), Hastinapur, Delhi (600 BC) Mohanjodaro, Sind (2500 BC) all show evidence of the existence of toilets and baths. But this practice of constructing toilets and baths in individual homes somehow disappeared and toilets were not considered a necessity as fields served the purpose. The provision of latrines and baths thus became an inessential feature in most houses in India largely due to cultural belief that place of defecation in the house or adjunct to it would defile the sanctity of the house. But over the years, toilets of different designs became popular. However, most of these designs could not stand the test of environmental sustainability, with the result, open defecation continued unabated. In a predominantly agricultural country, the prevalent practice is to defecate in the field away from home to return human excreta to the soil or near a flowing river or pond. The habit so cultivated continues to be a tradition and considered more convenient. Open defecation and bathing have long been part of morning ablutions, especially in rural areas, and urban slums. To meet the massive challenge of open defecation India needed low-cost, affordable, acceptable and eco-friendly toilets for people who could not afford to have western style toilets – a lavish loo at their homes. This, indeed, led to affordable public toilet revolution in the country. The civic bodies in India are now embarking on the mammoth task of promoting the construction of low-cost affordable toilets for low income groups, and simultaneously getting constructed pleasant public conveniences for the floating population in all cities and towns. The Government, in association with sanitation sector NGOs, is busy upgrading the existing public sanitation facilities that have remained insufficient and poorly managed on the one hand, and stepping up its efforts in promoting low-cost sanitation technologies on the other. This is not only to improve the quality of life but also to enhance the country's image overseas.

The technological option, most suitable for country like India (as also for several other countries in the developing world) is the one which provides a socially, culturally and environmentally acceptable level of service at the least economic cost. Both sewerage and septic tanks are not suitable for a developing country like India due to high cost and other limitations. Considering the vastness of the country with its divergent culture, social customs and attitudes, together with variations in climate, geological and hydrogeological conditions and low income of people, design of latrine which suits the need of rural and urban population raises several complex questions. In India, 1940 onwards, a dynamic search for a safe and economical alternative to sewerage and septic tank systems for the disposal of night soil suited to socio-cultural and economic needs to be started. Various low-cost latrines were tried, but they could not satisfy the Indian conditions and situations. It is at this point that Sulabh's cost-effective and eco-friendly sanitation technology was devised, developed and implemented. The eco-friendly twin-pit composting toilet with on-site human waste disposal system, as developed and designed by Sulabh International Social Service Organisation was considered to be the best available technological option to replace the bucket/dry privies. Popularly known as 'Sulabh Shauchalaya Technology', this eco-friendly, affordable and hygienic sanitation technology has been widely accepted by the people, the Government of India and several international organizations, like WHO, UNDP, HABITAT etc. as technically sound, affordable and easily available, even to the economically weaker sections of society.

Another significant development in the history of public sanitation in India took place when the concept of community toilets was operationalized, first by the local civic bodies in various towns and cities. Due to poor standards of cleanliness and day-to-day maintenance, these toilets did not receive widespread public acceptance, and were either abandoned or put into disuse. Again, the Sulabh International changed the concept of free public toilets, their design, infrastructure and ambience and replaced them with 'pay and use' community toilets, (approximately 7,500), operated and maintained by Sulabh. These toilets are better built, better maintained, better serviced, with facilities that good public toilets provide to their users in the countries of the developed world. The sanitation facilities (Sulabh Shauchalayas and Sulabh Community Toilet Complexes) provided by Sulabh International Social Service Organisation are being used by 10 million people daily. Hundreds of thousands of Sulabh Shauchalayas (twin-pit pour flush toilets) have been constructed all over the country in both rural and urban areas. Sulabh alone has constructed more than 1 million such Shauchalayas throughout the country. People have accepted them and are fully satisfied with their performance. The Government of India, State Governments, various national and international agencies like UNICEF, World Bank, UNDP, WHO, UNCHS etc. have accepted the design and are advocating its adoption in other developing countries in South-East Asia, Latin America, Africa etc. Impressed with the success of Sulabh programme, the Governments of South Africa, Sri Lanka, Nepal, Bhutan, Tanzania, Kenya and Afghanistan have sought the assistance of the organization to improve the sanitation status, environment and quality of life in their countries by following the Sulabh Model.

Sulabh's Approach to National Sanitation Challenge

Sulabh accepts an approach to sanitation challenge which is not only human - centred but also ecologically sustainable. It is one which is concerned with equity, the protection of environment, and the health of both the user and the general public. Its goal is to create socially, economically and ecologically sustainable systems. To reach this goal, Sulabh endorses three globally accepted principles which have been identified as critical to designing successful sanitation systems for the future : (1) equity within the sanitation sector which means that all segments of society, rich and poor alike, have access to safe, appropriate sanitation systems adapted to their needs and means; (2) health promotion and protection from disease which means that systems are capable of protecting people from excreta related disease transmission; and (3) protection of environment which means that future sanitation systems must neither pollute ecosystems nor deplete scarce resources. Programmes that fulfill these principles simultaneously will lead to long-term sustainability.

In its pan-India outreach of sanitary programmes and activities spanning over a period of 4o years, Sulabh has realized that the low-sanitation status of the majority of developing countries is wholly because of the fact that sanitation experts (policy planners, sanitation engineers) still cling to the notion that western sanitation technologies (sewerage and septic tank) offer the best (and perhaps only viable) options to the mammoth shortfalls in the domestic and public sanitation domains. While recognizing the importance of sewerage and septic tanks systems of sanitation, we do realise that the majority of the developing countries of Asia, Africa and Latin America, given their low income status, technology constraints and limitations of sanitation infrastructure building, cannot solely and wholly depend on sanitation technologies which have proved incapable of meeting the sanitation challenge the developed countries face and successfully confront. These 'big' technologies are suitable for the affluent countries, and not for resource starved countries of the developing world. These technologies are neither affordable, nor socially and culturally acceptable to many people, groups and communities, because of different social habits, cultural preferences, population profile, water availability constraints and so forth. The real good, viable, acceptable, affordable option rests in adopting low-cost eco-friendly and location specific technologies which are suitable and appropriate for local users. Hence, I have been repeatedly saying at all available opportunities and at different national and international fora that the developing world will not be able to meet the sanitation related Millennium Development Goals by 2015 or even beyond. What I suggest on the basis of our over four decades' experience is that low-cost sanitation technology, the kind of which Sulabh has devised, developed and operationalised in India, is the best for the developing countries.

Sulabh's toilet technology puts a premium on small scale and self sufficiency. In fact its simplicity and small scale makes it appropriate to meet the local needs. This technology has made enormous impact on the life of millions of low income people in India, especially those who have no toilet facilities at their homes. One evidence of the impact is that toilet facilities created by Sulabh are now being used by over 10 million people everyday, all throughout the country. The impact is still more on the minds of the people, the decision makers in the country, and at the world level, all realizing the benefits arising out of the application of Sulabh's sanitation technology and recognizing its suitability. There has been dramatic change in the physical environment of the towns and cities where Sulabh system is in operation. 240 towns in India have been made scavenging free where human excreta is not handled manually by scavengers and all dry and bucket privies have been converted into Sulabh toilets. Sulabh has so far constructed 12,00,000 (1.2 million) household pour-flush toilets (Sulabh Shauchalayas), besides constructing and maintaining 7,500 public toilet-cum-bath complexes. The system of operation of community toilets evolved by the organization has proved a boon in keeping the towns clean and in improving environment. This is a unique example of partnership of local authorities, non-governmental organization and the community. The provision of individual Sulabh type household toilets as well as Sulabh Community toilets-cum-bath complexes has minimized the indiscriminate open defecation, resulting in the improvement of environment and health of the community. Women and children are benefited because they were the worst sufferers of the lack of toilet facilities.

Chapter - VII

THE CONTRIBUTION OF SULABH INTERNATIONAL

Sulabh International Social Service Organisation (popularly known as Sulabh) is India's premier NGO which has been actively involved in sanitation since its inception in 1970. Its mission is to fulfil Mahatma Gandhi's unfinished task to restore human rights and dignity to the 'untouchables', while providing affordable sanitation facilities to masses through out the country. Its vision is to create a healthy and hygienic India, free from the practice of open defecation, indeed, an India free from faecal pollution of environment. To realize its mission and vision, Sulabh has initiated a nationwide movement, known as 'Sulabh Movement'. While Sulabh International Social Service Organisation is an institution to implement its programmes and objectives, the Sulabh Sanitation Movement is a combination of 'hardware' (technology) and 'software' (information, education and communication – IEC) components of sanitation.

The organisation was set up against a historical backdrop of 1930s when dynamic search was started for finding a safe and affordable alternative to sewerage and septic tank systems for the disposal of human excreta ideally suited to country's socio-cultural and economic conditions. Various low-cost sanitation options were tried without much success. It was in the post-Independence period that attempts were made to find ways and means to abolish scavenging. The Government of India set up various Committees and Commissions to improve the lot of scavengers.

Various State Governments, research institutions and non-governmental organisations set their minds to evolve a suitable system for replacing the bucket/dry privies, which could be adopted in the congested localities of urban areas. In spite of all these efforts, the programme launched did not achieve the desired results, as no economical and acceptable system could be evolved and no well-formulated programme for training and rehabilitation of the relieved scavengers was available.

It was in 1970 a voluntary social organisation known as Sulabh Shauchalaya Sansthan (now Sulabh International Social Service Organisation) was founded by the author at Patna (Bihar) to take up the liberation of scavengers as part of the passion which he had developed while serving the committee set up for the celebration of the Birth Centenary of Mahatma Gandhi.

The central idea behind the setting up of Sulabh Shauchalaya Sansthan was to bring about social and economic upgradation of about one million scavenger families, not through loud and populist slogans, but through a concrete programme of action backed and buttressed by viable technological alternatives and social and economic inputs. The idea was rooted in the belief that until the dry/bucket latrines exist, manual scavenging cannot be eradicated and, by implication, scavengers cannot be liberated from a demeaning and dehumanizing occupation of cleaning and carrying human excreta.

The search for a viable technological alternative to replace 'service latrines' requiring the services of scavengers for cleaning and carrying human excreta was not an easy one. Though there existed various alternative models for safe disposal of human excreta, namely sewage and septic tank, they were not found suitable in view of the country's socio-cultural and economic conditions. The search for low-cost sanitation methods and their widespread adaptability was difficult because these methods had earlier been tried in different parts of the country without much success. And since a better and low-cost method was not available, manual scavenging could not be abolished and scavengers continued to be tied up with their unclean and socially degrading hereditary occupation.

After years of study and research on various technological alternatives to bucket/dry latrines, a solution emerged. The solution was pourflush waterseal latrine with two leach pits for on-site disposal of human excreta, popularly known as Sulabh Shauchalaya, meaning a convenient toilet. The innovation was a breakthrough in the Government of India's programme for eradication of manual scavenging, because the system, as devised, proved affordable, technically sound, acceptable to the masses and easy enough to be set up with available local facilities.

While the idea of Sulabh Shauchalayas was an innovation in the whole range of low-cost sanitation technologies, it took several years to gain ground and get the official approval for its countrywide implementation. How it received the Government's support is a long history. The pilot project in a municipal town of Arrah in Bihar State proved so successful that the idea soon caught the attention of appropriate authorities in different parts of the country. The word that spread about the success of the novel experiment in Bihar, soon caught the attention of other State Governments in the country, including the Government of India.

The WHO, with the assistance of the Ministry of Works and Housing (now Urban Development), Government of India and UNICEF examined the Sulabh's low-cost sanitation technology. The experts found conclusive evidence to opine that two-pit pourflush waterseal latrine with on-site sanitation, known as Sulabh Shauchalaya is the most appropriate low-cost sanitation technology having technical, social and cultural implication of far-reaching significance to deal with the problem of manual scavenging and the liberation of scavengers from a demeaning and dehumanizing occupation.

The appreciation of the scheme by the relevant national and international bodies (Government of India, WHO, UNDP and UNICEF) soon led to the expansion of organisation's programmes and activities in other States. When Sulabh's network widened and was about to transcend the confines of Bihar State, the name of the organisation was changed from Sulabh Shauchalaya Sansthan, to Sulabh International.

In less than a decade, Sulabh International carried forward its journey to cover nineteen States and two Union Territories in the country. This, indeed, was a giant leap forward for an organisation which initially had a local and regional base. The organisation has now become truly national in its outreach. The suffix 'International' is justified in as much as its idea and methodology of action has significance for developing countries of Asia, Africa and Latin America, where the problem of environmental sanitation is exceedingly complex and formidable.

The organisation now has its branches in 1075 districts/ towns in 26 States and 3 Union Territories. It has a massive manpower base of over 60 thousand associate members and volunteers comprising social reformers, administrators, planners, engineers, scientists, social scientists, architects, medical doctors, sociologists, economists and social workers.

Objectives

The growth of the organisation soon led to an expansion of its programmes and activities to cover several broad areas of social transformation and social development. In view of its expanded social mandate, the organisation's name was again changed from Sulabh International, it is now named as Sulabh International Social Service Organisation with following objectives :

1. Liberation of Scavengers
   - to liberate scavengers from unhealthy and sub-human occupation of carrying human excreta as head-load by converting bucket latrines into pour flush toilets, popularly known as Sulabh Shauchalayas;
   - to rehabilitate the relieved scavengers in other occupations like general cleaning, store keeping, gardening, messengers, cleaners on garbage disposal vehicles etc.;
   - to train the sons and daughters of the scavengers in various vocations like shorthand, typing, leather work, plumbing and training as mechanics and electricians, T.V. & radio repairing, computer training etc. so that they could be employed or self employed in the above fields;
   - to help the scavengers to participate in the socio-religious functions as equals in the society like temple entry, worship, marriages etc.;
   - to persuade the well-to-do families to adopt at least one scavenger family so that they could be brought socially and economically at par with the other members of the society;
   - to establish information and employment exchange centers to help the scavengers in getting employment either in the government or non-government organizations;
   - to open English medium schools to provide quality education to the sons and daughters of the scavengers along with others;
   - to help scavengers build their houses away from the slums so that they could be taken out from the stinking environment.
2. Prevention of environmental pollution and improvement of health, hygiene and ecology.

- to educate people not to defecate in the open, in lanes, bye-lanes, parks, streets, on road side or on rail routes etc.
- to motivate people to get bucket latrines converted to pour-flush toilet (Sulabh Shauchalayas), and also to set up toilet facility in houses where none exists.
- to provide community toilets with bathing, washing and urinal facilities on 'pay and use' basis in slums and for pavement dwellers, rickshaw pullers and the floating population, and also for those communities in whose houses latrines cannot be constructed due to space constraint;
- to motivate people in the rural areas to have the toilet facilities in their houses to save women, children and older people from the inconvenience of going out for defecation in the open and to protect them from the diseases like hookworm, ringworm and cholera etc., besides protecting the honour, modesty and dignity of women folk.
- to persuade people to plant trees around the public toilets for the protection of the environment.

3. To harness non-conventional energy sources from human waste and save fuel and forest.
   - to produce biogas from human excreta to be used either for cooking, preparing hot water for bathing, washing and warming bodies during winter season and to light mantle lamps; and
   - to convert biogas into electrical energy with the help of the dual-fuel engine to be used for lighting the streets, community complexes and houses.
4. To procure manure form the Sulabh Shauchalayas and Sulabh Complexes and use it to raise farm productivity;
5. To assist the community in building low cost houses;

6. To assist the Government in providing safe and sustainable sanitation facilities to the people in rural and urban areas;
7. To promote job oriented education, primary health care and integrated rural development;
8. To form groups of people from all walks of life throughout the country to build public opinion against social evils which afflict the economically poor and neglected sections of the society;
9. To create new employment avenues by training 'change agents' for rural and urban development;
10. To promote consultancy, research and development in technical and social fields; and
11. To undertake diffusion of innovations, education, motivation and awareness through mass communication.

Sulabh Shauchalaya

Sulabh Shauchalaya consists of pan with a steep slope of 25^0 to 30^0 and trap with 20 mm waterseal needing 1.5 to 2 liters of water for flushing. The excreta is carried into leach pits through pipes or covered drain: one pit is to be used at a time. The liquid infiltrates and gases disperse into the soil through the holes in the pit lining. When one pit is full, the excreta is diverted to the second pit. In about 1.5 to 2 years, the contents of the pit already filled get digested and become safe for handling. The pit can then be conveniently emptied and made ready to be put back unto use after the second pit is full. Thus, the two pits can be used alternately and permanently. The sludge of the pit is a good manure for use in horticulture/agriculture. The cost of emptying the pit can be met from the cost of manure available.

Sulabh Shauchalaya fulfils all the seven conditions for a sanitary latrine laid down by Wager and Lanoix in their book

– *Excreta Disposal for Rural Areas and Small Communities (1958).* These are as follows :

i) The surface soil should not be contaminated

ii) There should be no contamination of ground water

iii) There should be no contamination of surface water

iv) Excreta should not be accessible to flies or animals

v) There should be no manual handling of excreta

vi) There should be no bad odour or unsightly conditions.

vii) The method used should be simple and inexpensive in construction and operation.

Advantages of Sulabh Toilets

Key advantages of Sulabh's two-pit pour flush composting toilets are :-

- Hygienically and technically appropriate, and socio-culturally acceptable
- Affordable and easy to construct with locally available materials.
- Design and specification can be modified to suit householder's needs and affordability.
- Eliminates mosquito, insect and fly breeding.
- Can be constructed in different physical, geological and hydrogeological conditions.
- Free from health hazards and does not pollute surface or ground water, if proper precautions and safeguards are taken during construction.
- Can be constructed on upper floors of houses.
- Pits are generally designed for 3-years' desludging interval, but if desired, they can be designed for longer periods or can be reduced even to two years.

- Maintenance is easy, simple and costs very little.
- Needs only 1.5 to 2 litres of water for flushing, while conventional flush toilet needs 12 to 14 litres of water.
- Needs less space than septic tank linked toilet system.
- Does not need the services of scavengers for cleaning the pits or disposal of sludge. This can be done by the householders.
- Makes available rich fertilizer and soil conditioner.
- Can be easily connected to sewers when introduced in the area.
- A low volume flushing cistern could be attached to avoid pour flushing.

To take the model to the people at large and make it acceptable to them, Sulabh International's initial problem was that the new design of privies was to be constructed in households and used by individual families. The households were to adopt a new technology, switch over to a new practice, cultivate new habits in place of age old practice and social habits. Thus the main thrust aimed at changing the attitude of the people towards better sanitation and motivating them to have Sulabh Shauchalaya in their households and use it. It was obvious that a public agency in charge of such project could not attain the project objective. Sulabh's appearance in the field of sanitation made a breakthrough in the Government of India's efforts for eradication of scavenging.

Sulabh International through its trained social workers started motivating the people in accepting the new technology and educating them to properly use and maintain their own latrines. It was instrumental in motivating the Government of Bihar to provide 50% grant and 50% loan to the beneficiaries through the local authorities for conversion programme. To give a boost to the programme, the organisation started educating and motivating the households to accept the new technology and assisted them in getting the applications processed, and loan agreement executed and sanctioned on their behalf. This procedure met with great success. The reason for this was that the beneficiaries got all the facilities at their door-steps and the entire work was carried out as a turn key job providing a guarantee for 5 years.

After studying the work done in Bihar, the WHO with assistance of the Ministry of Works & Housing (now Urban Development), Government of India in collaboration with Sulabh International and UNICEF convened the first National Seminar on conversion of bucket latrines in to sanitary water seal latrines in May 1978 at Patna. The Seminar concluded that two-pit pour-flush water-seal latrine with on-site sanitation, popularly known as Sulabh Shauchalaya, is the most appropriate low cost technology to be introduced on a large scale in the urban areas of the country, as the system meets the technical, social and cultural needs of India.

The Bihar programme was an eye-opener to the Government of India where by about 50,000 dry/bucket privies were converted into Sulabh Shauchalayas in the most thickly populated towns of Bihar. The Government appreciated the importance of low cost sanitation in liberating the scavengers and in improving the present insanitary condition by observing the success achieved by Sulabh International in Bihar.

In 1979 when the UNDP/World Bank took up the proposal for demonstrating the installation of low-cost sanitation on a worldwide basis, based on the experience of Sulabh International, the Government of India decided that UNDP/World Bank should take up the Feasibility Study on low cost sanitation in India in 211 towns spread over in 21 States and Union Territories. Several studies were carried out by the UNDP/World Bank of the Bihar Programme. The studies included the evaluation of the programme, investigation of various technical features and the assessment of pollution from the existing leach pits. The convincing data based on actual work done in varied physical, hydro-geological, social, cultural and economic settings in Bihar provided useful material to draft the Feasibility Reports on low cost sanitation. Thus, it was Sulabh International's successful achievement that showed the path for adoption of low cost sanitation system on a large scale, which was finally accepted all over India. The system developed by Sulabh is responsible for the emancipation of scavengers from the demeaning task of manually handling of human excreta and has direct impact on the improvement of community health and environment.

The author convinced the Ministry of Welfare, Government of India to provide 50% subsidy to boost the latrine conversion programme in different States by citing the success of low-cost sanitation programme in Bihar.

During the Sixth Five Year Plan (1980-85), the Ministry of Welfare, Government of India started financing projects with 50% subsidy by the Government of India and the balance to be met by the State Government as loan/grant for conversion of dry/bucket latrines into Sulabh Shauchalayas to make towns scavenging free.

Sulabh International has so far converted/constructed 1.2 million toilets/privies into Sulabh Shauchalayas and constructed new ones in houses where none existed before in 1247 cities/towns in 25 States and 4 Union Territories. Thus nearly 1,20,000 scavengers have been liberated from the demeaning task of manual handling of human excreta.

Type of Sulabh Shauchalaya

There are three types of Sulabh Shauchalayas : (1) Household Sulabh Shauchalayas (Individual toilets); (2) Sulabh Shauchalayas for Public use (Community toilets); (3) Sulabh Shauchalayas with biogas plants (Public toilets connected with biogas plants).

Household Sulabh Shauchalayas

Household Sulabh Shauchalayas (individual toilets) serve the need of a single family. Sulabh International has developed several designs of these toilets according to the need and cost affordability of the people of various socio-economic strata in different parts of the country. The cost of each design varies, but always remains within the reach of people in the lower and middle economic strata. The cost of Sulabh flush composting toilet varies widely to suit people of every stratum. The cost ranges from US$10 to US$500 per unit. It depends upon materials of construction of pits and seats as well as of superstructure.

These toilets have found wider public acceptability throughout the country. The Government of India, State Governments, various national and international agencies like UNICEF, World Bank, UNDP, WHO, United Nations Centre for Human Settlement (UNCHS) etc. have accepted the design and are advocating and providing financial assistance for the construction of Sulabh type individual toilets in houses without them. Conversion of dry latrines into Sulabh Shauchalayas has also been taken up on a large-scale in both urban and rural areas all over the country.

Public Conveniences – Sulabh Community Toilet Complexes

To supplement the household sanitation system, Sulabh conceived the idea of construction and maintenance of public toilet complexes in the urban areas mainly for the pavement dwellers, the floating population and the local population which can not afford to have latrines in their houses for lack of space or for economic reasons. It is generally observed that public conveniences put up by the civic authorities soon fall invariably out of use simply because of lack of proper cleaning and maintenance. These facilities get neglected and in a shortwhile become the source of health hazard and people hardly use them.

The year 1974 was a landmark in the history of sanitation when Sulabh International entered the field with its own design of construction of Sulabh Toilet Complexes and operating and maintaining them on 'pay and use' basis charging a nominal amount from the users of latrine and bath, the use of urinals being free. For women, children and handicapped, it is entirely free. These complexes became immediately popular and created public demand. The design of Sulabh Complexes has special features. There are separate enclosures for men and women. Besides, toilet complexes also have urinals and bath with continuous supply of water, gardens and pathways inside for aesthetic and hygienic purposes. Special squatting pans are provided for children. The complexes are kept properly lighted during night. Sufficient opening to the sky allows good ventilation and direct entry of sunlight. Soap powder is provided to the users for washing their hands and developing a sense of cleanliness. Green trees and shrubs are planted to create shade and to make the complexes pleasing to the eye.

The high standard of cleanliness and maintenance of Sulabh Complexes are their main attraction. They are kept open day and night with round the clock service of attendants and supervisors for operational management and to keep the complexes neat and clean. The funds generated by the collections from these complexes are used for their routine and long-term maintenance and to meet the organizational expenses. Thus it relieves the local authorities from day-to-day maintenance.

Apart from toilet and bathing facilities, Sulabh International has started providing medical aid, public telephone and left luggage facilities in some of the complexes. It will be a feature in all the Sulabh community toilets in the near future. An ambulance car will also be made available where possible.

The Sulabh Complexes with a "Pay and Use" system have now found acceptance throughout the length and breadth of the country and have earned international appreciation also. Till now more than 7,500 such public toilets have been constructed and are being maintained by Sulabh. The biggest public toilet of Sulabh has been constructed at Shirdi, in the district of Nasik, in the State of Maharashtra, having 120 WCs, 100 bathrooms, 28 special toilets (separate for ladies and gents) and 5,000 lockers for the convenience of the pilgrims. The scheme has been welcomed and accepted both by the people and the government. It has restored human dignity because the people who were forced to resort to open defecation got the opportunity of using a sanitary latrine. Availability of bathing facility is another boon to the economically weaker sections. These measures have resulted in better health and environment, hence higher productivity. These complexes have been of particular help and benefit to women. The accessibility to a toilet and bath is a boon to them, and takes care of their safety and ensures their modesty.

Sulabh toilet complexes are located in public places, bus stands, hospitals, market places and slums. For the construction, operation and maintenance of these complexes, Sulabh plays the role of a catalyst and partner between official agencies and the users of the complexes. The system of operation and maintenance of community toilets has proved a boon to the local bodies in their endeavour to keep the town clean and improve environment. This is a unique example of partnership between local authorities, a non-governmental organisation and the community. Sulabh's concept of such 'pay and use' toilets is an important contribution to the promotion of environmental sanitation in the country.

Public Toilets Linked with Biogas Plants

The third types of Sulabh Shauchalayas are those that are linked with biogas plants. The installation of biogas plants has been possible in public toilets constructed and maintained by Sulabh International, which have a substantially large number of users every day. Sulabh International is the first to generate biogas from human excreta on large-scale at public toilets.

Sulabh is a pioneering organisation in India in the field of biogas generation from community toilets linked biogas plants. The organisation started its effort in 1982 when the first such biogas plant was set up at Patna. Keeping in view the obnoxious smell, psychological barrier with human excreta, quantity of water used each time and the technical problems related with feed materials, there was a gradual improvement in the design of biogas plant. The design developed by the organisation does not require manual handling of the human excreta; the biogas plant is underground and the flow of excreta into digester is under gravity. After anaerobic fermentation of human excreta, biogas is produced. The effluent that comes out of the digester through the outlet chamber is stored in soak pits – where sewer line is not available. If available, it is connected to sewer. The effluent and sludge is used for irrigation and as manure. 192 community complexes have been connected with biogas plants.

The biogas produced from public toilet complexes is being utilized for four different purposes e.g. cooking, lighting, electricity generation and body warming during winter. Biogas can be efficiently used for cooking purposes. It burns with blue flame without any soot and odour. In rural areas where cowdung cake and wood are the only sources of fuel, biogas from human waste is a boon. Based on its R&D efforts, the organisation has developed a new and more effective design of biogas plants that has been approved by the Ministry of Non-Conventional Energy Sources (MNES), Government of India. The Ministry is providing subsidy for the construction of such biogas plants through its different state agencies. The Ministry has approved the design of

Sulabh and has financed the installation of 140 biogas plants of 35 to 60 cum capacity in different parts of the country.

Endeavours in Social Transformation

In its long journey of three and a half decade, Sulabh International has transcended its confines of low-cost sanitation and has ventured to work in other significant areas of social transformation. The organisation now dons a new role-outfit of the dimension of a wider social movement. The Sulabh Movement as it is being currently referred to in sociological parlance, endeavours in the following thrust areas of social transformation :

- Liberation of Scavengers : An endeavour at restoring human dignity to the underprivileged;
- Training and rehabilitation of liberated scavengers and their wards;
- Social transformation and upgradation of untouchables (scavengers);
- Social transformation through social engineering: An attempt of bringing high caste people into sanitation sector;
- Adoption (Befriending) of the scavengers families : A significant measure of social mainstreaming; and
- Sulabh movement's strategy of rural reconstruction and human resource development : Again a step towards social transformation.

A brief description of these thrust areas of social transformation is as follows :

Liberation of Scavengers

One of the more prominent and nationally and internationally acclaimed dimensions of Sulabh Movement is the liberation of scavengers from the demeaning and dishonouring task of cleaning dry latrines and carrying human excreta.

Using Sulabh technology, bucket latrines were converted into pourflush toilets, thus paving the way for liberation of scavengers from the centuries-old sub-human practice of carrying human excreta. Sulabh's idea was a breakthrough in the Government of India's attempt at the eradication of scavenging, as the system developed by Sulabh International is inexpensive, acceptable to masses and could be implemented with the available local resources.

Following the rapid adoption of the scavenging-free Sulabh technology of pourflush toilet system, 640 towns of India have been freed from scavenging and 1,20,000 scavengers have been relieved from the demeaning practice of manual handling of human waste and are now engaged in various professions. This was achieved by converting around 1.2 million bucket latrines into Sulabh toilets in 40 years. No other organisation, except Sulabh could achieve such a huge success in the field of sanitation in such a short time. Its achievement is even more noteworthy, considering the fact that it had to work with meager resources and being a voluntary non-profit organisation. This is the only NGO in the country which is fully self-sufficient and does not receive any grant or funds from national and international organisations.

Sulabh shares the credit of persuading the Ministry of Welfare (Government of India) to undertake the programme of liberation of scavengers under the protection of Civil Rights Act (1955) in the year 1980. The campaign was so successful that the central Government later decided to get converted some six million bucket latrines into Sulabh Shauchalayas and provide individual Sulabh toilets in four million houses having no toilet facilities at all. All this was aimed at abolishing scavenging within five years.

Training and Rehabilitation of Liberated Scavengers and Their Wards

Though the introduction of low-cost sanitation

technology had resulted in the liberation of scavengers who were earlier engaged in disposing of human excreta, the employment problem of the sons, daughters and the wards of the liberated scavengers remained intact. It was, therefore, found necessary to make arrangements for the training of the wards of liberated scavengers, so that they could be rehabilitated in different occupations and brought at par with other fellow citizens in terms of their socio-economic status.

Sulabh International formulated a plan of training the liberated scavengers and their wards and set up training Institutes in three States and one Union Territory – Bihar, Maharashtra, Rajasthan and Delhi with the assistance of the Ministry of Welfare, Government of India. Some 7,000 scavengers have been trained at various vocational Training Institutes and Centres run by Sulabh in various cities in India. Most of these trained scavengers have taken up economically rewarding occupations. Many of them have been employed in the formal and informal sectors of business and industry.

Sulabh International is the only NGO which has taken up the twin operations of liberation of scavengers by conversion of dry/bucket privies and training and rehabilitation of liberated scavengers and their wards. The organisation has already formulated a National Plan for the training of liberated scavengers and their wards. The strategy envisages setting up several such Training Institutes in different parts of the country and the detailed proposals in this regard have already been submitted to the Government of India (Ministry of Welfare) and respective State Governments. Sulabh International proposes to train 35,700 wards of the scavengers in 10 years from now.

Sulabh Public School (Education of Scavengers' Children)

Education holds the key to any major change and development and is essential to improving the condition of the traditionally oppressed Scheduled Caste communities.

Having this objective in view, Sulabh has endeavoured at imparting good education to the scavengers' children in the school-going age. It has set up an English Medium School, the first of its kind, in Delhi to provide 'quality' education to the children belonging to the socio-economically weaker sections of society. The special features that merit attention lie in an attempt to narrow down the educational gap between the rich and the poor, especially between the children of the scavengers' families and the others. The Sulabh School can be matched with any other school for the children of the elite. The significance of the school lies in not being an elite educational institution, which are too many in the country; it lies in breaking the tradition of setting up the government schools that impart inferior education to the children of the weaker sections of society. The English medium schools in India are the monopoly of the rich, and the kids of these schools develop proficiency English and capture plum jobs on that basis. Another peculiar feature of these schools is that the parents and guardians of the children seeking admission in these schools must know how to converse in English language at the time of their interview. Scavenger parents are doubly handicapped; on the one hand they cannot afford the prohibitive cost of education, and on the other, they cannot converse in English at the time of interview.

Viewing the complexity of the situation, the Sulabh School put a premium on English as the main medium of instruction. But, at the same time, the importance of other languages is not underrated. Equal, if not more emphasis is given to Sanskrit, the language in which the Brahmins were supposed to be adept. It has, therefore, been made compulsory for children in the School to acquire proficiency not only in English, but also in Sanskrit to break the Brahmin domination over the latter. Besides these two languages, the school lays equal stress on Hindi, the national language.

The Sulabh School in Delhi is a pace setter. It intends to

equip the traditionally disadvantaged groups of Indian society with the instrument of education (a real social leveler) for their upward mobility in the changing social milieu. The school is open to the children of non-scavenger's families to avoid segregation. This has been done with a definite purpose of enabling the children of scavenger's families to mix freely with the children of the higher castes and rub shoulders with them as equals in society.

The school aims at offering education at least upto higher secondary level. Apart from providing free education, the school also provides for free conveyance and two pairs of complete uniform to each student in a year. Healthy surroundings, better teaching aids, a smaller teacher-student ratio and better organised extra-curricular activities are the special features of the school. Another distinguishing mark of the school is that it makes arrangement for the training of children in different vocations, such as typing, computer-handling and office management, right from the fourth standard. The School is recognized by the Directorate of Education, Government of Delhi and provides education upto Tenth Standard. Sulabh has a plan to set up such schools in all States.

Social Transformation and Upgradation of Scavengers

Sulabh Movement is concerned with and committed to accelerating the process of social transformation in respect of untouchables (especially the scavengers). Social engineering or ameliorative social action is the hall-mark of Sulabh Movement's social service dimension. The Movement hates all forms of social discrimination against the untouchables in general and scavengers in particular. The movement discards the thesis that the members of certain castes/communities would remain perpetually shackled to certain unclean occupations, or for that matter, remain tied up with their degraded traditional occupation. It is because of this abhorrence for caste-based discrimination against untouchables and especially against the scavengers, the

Sulabh Movement is engaged in the struggle for the liberation of scavengers. The ceaseless efforts of the Movement in restoring the self-respect and dignity to scavengers have paid rich dividends and Sulabh has succeeded in freeing thousands of scavengers from stigma and social bondage.

Not content with the liberation efforts alone, Sulabh Movement has embarked upon a novel strategy of scavengers' social upgradation. The motivating idea is that liberation from an unclean occupation is not enough; what is necessary is the action aimed at the assimilation of untouchables in the mainstream of social life. Liberation alone is not enough, because social problems like untouchability and social boycott continue to plague the system. The liberated scavengers are kept off by certain high-caste Hindus who do not accept the untouchables as even near-equals.

The social upgradation strategy of Sulabh Movement entails the raising of social status of scavengers in order to enable them to live with high-caste Hindus. The social upgradation of scavengers is an act of social intervention to bring about the desired change in the Hindu orthodoxy. The Sulabh Movement, therefore, tries to remove distortions and attempts to establish a discrimination free society. It does not seek to pitch Harijans against non-Harijans, because, doing that would expose Harijans to greater travails. The Sulabh Movement propagates the belief that time is not for confrontation between groups and religions, for conciliation alone would produce order out of highly inflammable religious-political tumult that the country is passing through.

Social Transformation Through Social Engineering

Social transformation through social engineering is another important dimension of Sulabh Movement. Before the Movement embarked upon an ambitious country-wide programme for the liberation of scavengers, toilets and other sanitation facilities were taken care of only by Mehtars, the lowest among the low in the caste-line up. In India nobody could have thought that one day toilets would be looked after and managed also by high-caste Hindus, including Brahmins. Now, the Sulabh Movement is thronged by caste-Hindus, majority of them belonging to so-called twice-born, especially the Brahmin. In this way, Sulabh Movement has brought about an appreciable change in the attitude of the people, that the so-called dirty jobs are not to be done by the lower caste people alone. Unlike earlier times the management of Sulabh Toilet and Bath Complexes is being done by the people of all castes, who have willingly joined the Sulabh Movement as 'volunteers'.

Adoption (Befriending) of the Scavenger Families

A very significant dimension of Sulabh Movement is its recent programme of persuading the well-placed high-caste Hindus to adopt one scavenger family each. This programme is consistent with its overall strategy of social absorption and mainstreaming of scavengers. Hundreds and thousands of letters and appeals have been sent to the distinguished persons in the country, requesting them to adopt one scavenger family each. The response to the appeal has been fairly encouraging. Already, more than 10,000 persons belonging to elite groups have expressed their willingness to 'adopt' one scavenger family each.

Sulabh's social upgradation programme for the social mainstreaming (social integration) of scavengers has been hailed as a landmark. Social upgradation of scavengers is a novel concept. This answers the question whether after liberation, training and education, scavengers and their children will enjoy social respectability and status in society at par with the so-called upper caste people in India. By linking scavenger families to persons committed to their welfare, the intention is not to provide them with status, but to inspire others to shed social prejudices and embrace scavengers as their equals. Social upgradation programme is, thus, an attempt to undo the social injustice done to the members of the scavenging caste. The social upgradation programme is necessary because scavenging is not entirely a problem of money, technology or law. It is also a cultural and attitudinal problem which begins in the minds of men. Hence through persuasion and pressure, the 'Movement' seeks to link defenseless scavengers to eminent people so that they can fight the battle for social justice with the support of the 'adopter' families whose most important obligation should be to work in their own quiet ways to end this inhuman practice and save society from being split into another hostile class. The Indian society is already too deeply divided to divide again. The question is not wholly of equity, but also of our survival as a united nation.

The social adoption of scavenger families, as being spearheaded by Sulabh Movement, does not involve any legal obligation; the adopters are only persuaded to befriend the scavenger families, taking food with them and occasionally inviting them at social gatherings. All these token gestures raise the social status of scavengers and help them get absorbed into the mainstream of social life.

Sulabh WASH Campaign

Recently Sulabh has launched WASH campaign as part of its commitment to provide safe sanitation. Sulabh has prepared its volunteer force to reach out to urban slums, squatter colonies and remote villages to approach women to inform them about the key issues and central messages of WASH. Initially Sulabh imparted training to 750 women volunteers from 12 urban slums in Delhi regarding the various aspects of health and hygiene and safe drinking water. These volunteers acting as lead persons undertook door-to-door canvassing on safe drinking water, health and hygiene. So far more than 12,000 persons, mostly women, have been trained. Sulabh has printed on a massive scale the posters, pamphlets, banners with attractive messages on health, sanitation and hygiene and distributed them in different slums and other areas. Under its aegis and with the support of Government of India, Sulabh started SulabhENVIS, News Letter in order to propagate the information on water, sanitation etc. A wide range of communication materials for local adaptation and application has been prepared for free distribution in selected impoverished neighborhoods in both cities and villages. Women are targeted as the principal beneficiaries of the Sulabh's WASH campaign. To exemplify its concern for women, Sulabh's WASH Campaign was kick started by a well-known female cine star as proof of Sulabh's focus on women in its unique and nationwide WASH Campaign. Under its WASH campaign Sulabh has setup three 'Sulabh School Sanitation Clubs' in Orissa in the month of July 2004 and is currently engaged in the setting up of similar clubs in West Bengal, Andhra Pradesh, Jammu and Kashmir, and Maharashtra. Sulabh's WASH Campaign is readying itself to improve the sanitation and hygiene behaviour of people living in slums, villages and other selected unhygienic urban localities / settlements by creating awareness towards the use of clean water and sanitation.

Sulabh International Academy of Environmental Sanitation

Sulabh International Academy of Environmental Sanitation, erstwhile known as Sulabh International Institute of Technical Research and Training (SIITRAT) was started in 1984 with the objective of carrying out applied and fundamental research as well as training and consultancy work on various issues related to social, environmental and technical developments in the country. During the last two decades SIITRAT has carried out significant researches and consultancy projects on sustainable technologies like biogas plants, duckweed ponds, materials and designs of pan and trap of toilet, waste water disposal and treatment, solid waste management, bio-medical waste management etc. Some of the important technology-centred research projects carried out by SIITRAT that need special mention are as under :

I. **Duckweed Technology for Effluent Treatment**: The use of this technology for treatment of waste water has been tried out earlier in other countries like China, Bangladesh in rural and peri-urban areas and not in urban areas. Sulabh took the initiative to introduce duckweed for treatment of waste water in urban areas In Delhi, a pilot project was taken up which has shown that through duck-weed, waste water containing BOD 250 mg/l and COD 600 mg/l can be treated and it becomes pure enough for pisciculture. Duckweed based waste water treatment units have been installed in Orissa, Bengal, Punjab and Haryana. It is proved by now that small towns having a population up to one hundred thousand do not need any conventional treatment plant for treatment of waste water. Rather, through the use of duckweed, the waste water can be treated and can generate revenue through pisciculture. This technology can help the unemployed youth in rural areas grow fish and earn money.

II. **Water Hyacinth Technology**: The water hyacinth too can be used for treating the waste water as well as for increasing the production in a biogas plant based on cow dung or human excreta. The water hyacinth should be dried and powdered to be used in the biogas plant mixed with a little quantity of cow dung. The use of water hyacinth enhances the production of gas 3-4 times. So even with one cow and one calf a family may get bio-gas for 8 -10 hrs. From experience it was found that most of the biogas plants were not functioning properly in India because of improper feed.

III. **The Economy of Water:** The use of water in both the technologies of Sulabh Shauchalaya in individual houses or public toilets requires only 2 litres of water per use to flush; whereas in conventional system it requires minimum 10 litres of water per use for flushing. Thus the most precious thing – the potable water of around 6000 MGD – can be saved if calculated with the use of 700 million people.

IV. **Biofertilizer**: In Sulabh Shauchalaya technology, 40 kg of fertilizer is obtained per annum per person. If it is calculated for 700 million people @ 40 kg. the production of manure roughly works out to 28 Million tonnes per annum, the cost of which is roughly 15,000 crore of rupees. So while on the one hand it can help minimise the use of chemical fertilizers, on the other, it can raise agricultural productivity. In the world today there is demand for use of biofertilizer in agriculture and horticulture. It has been found that traces of chemical fertilizers remain in the food that people consume which ultimately find their way to the body and affect adversely the health system. So this technology is also of great value in the field of production of biofertilizers. Since the water discharged from the public toilets also contains phosphate, nitrogen and potash, it is very beneficial and could also be garnered in large quantities if the biogas technology is used in all the public toilets, housing colonies and high-rise buildings to be built in the future.

V. **Biofertilizer from Human Urine**: Sulabh is experimenting to obtain fertilizer from urine. From the preliminary studies it has been found that the use of human urine enhances agricultural yield more than from the use of other fertilizers. It has been found that human urine can be used as an alternative to chemical fertilizers to reduce pollution in water and damaging effects on soil and help control other environmental hazards caused due to the use of chemical fertilizers. The use of human urine in place of chemical fertilizers reduces the rate and amount of chemical fertilizer runoff into the groundwater and surrounding water ways.

Manure is essential to the soil to increase its fertility. Human urine can be an excellent high nitrogen liquid fertilizer for most plants. Fresh urine which contains nitrogen mainly in the form of urea is a good fertilizer. It can provide enough nitrogen, phosphorus and potassium and is almost free of heavy metals. When urine is used as fertilizer, the gardening costs reduce, compared to those incurred in use of chemical fertilizers.

The main aim of the Sulabh's R&D on human urine is to assess the nutrient availability in urine; to stabilize pH of human urine and to prevent loss of ammonia; to find out the best option of using urine as fertilizer; to find out the effect of urine on growth of plant when used as fertilizer; to compare the growth of plant using human urine against chemical fertilizers; to assess the feasibility of the use of human urine for making compost and also to suggest decreasing the pollution level in water bodies due to use of urine. On decrease of pollution in water bodies the cost of treatment of wastewater will be ultimately reduced.

Today almost 10 million people are being benefitted by the various Sulabh technologies mentioned above. Apart from the tremendous beneficial impact on community health, it has also resulted in socio-cultural development and economic benefits. In terms of saving of water alone, it is estimated that about 80 to 100 million litres of water is saved everyday by the people using Sulabh Shauchalayas. Similarly at the rate of 40 kg of bio fertilizer (NPK) produced per person per annum, the potential of resource recycling is at once evident.

VI. **Sulabh Thermophilic Aerobic Composter (STAC):** Sulabh has developed a new technology – Sulabh Thermophilic Aerobic Composter (STAC) that requires only 8 to 10 days to make compost from any biodegradable wastes without any manual handling during composting. It is based on thermophilic aerobic method. The technology does not require recurring expenditure. The plant is G.I. sheet made having double wall filled with glass wool, partitioned with perforated sheet into three chambers. After biodegradation, liquid is collected in bottom chamber that can be easily taken out and used for agricultural/ horticultural purposes. Manure that contains 30-35% moisture can be directly used for agriculture/landfill purposes or it can be dried, granulated and stored till further use. The practical utilities of this technology are: (i) organic solid waste can be efficiently converted into manure and soil conditioner having direct/indirect economic return, (ii) it can control diseases transmitted from wastes, as at high temperature pathogens are eliminated from the waste, (iii) due to reduction in volume, carriage cost of wastes to disposal site as well as area needed for landfills will be drastically reduced, and (iv) spread of weeds from wastes will also be controlled. This technology is more suited for rural areas as its byproducts (compost) can be readily used for agricultural purposes, and it reduces health hazards.

SIITRAT has a well-equipped laboratory with testing facilities and also for undertaking research and development in waste water treatment methods, low-cost sanitation technologies, development and improvement of biogas digestor system etc. One of its achievements has been testing, on the recommendation of the Delhi Pollution Control Board, the samples from effluent treatment plants of various industries in Delhi and providing certificates about quality of the effluent discharged. This has been done in respect of more than 400 samples so far.

Under a project funded by the European Union and in collaboration with the Bremen Overseas Research and Development Agency (BORDA), Sulabh has developed a Low Maintenance Wastewater Treatment System

(LOMWATS), under which as many as ten such treatment plants have been implemented and it is in the process of dissemination of the technology all over India.

A few other important Research and Development and Consultancy Projects undertaken by SIITRAT are mentioned below :

Waste Water Management:

- Prevention of pollution of the river Yamuna.
- Low Maintenance and Decentralized Waste Treatment System.
- Contingent evaluation survey for the Ganga Action Plan (users benefit survey), Haridwar, U.P.
- Consultancy for the preparation of project report on water supply, sewerage, storm water drainage and solid waste management in the proposed Ranka town, a satellite town of Gangtok, Sikkim.
- Duckweed based waste water treatment system associated with pisciculture in small and medium towns with economic return and also demonstration-cum-study project for economic development and employment generation in rural areas.
- Dal Lake conservation project under the National lake conservation plan.

Solid Waste Management:

- Hospital Waste Management in Ludhiana town, Punjab.
- Occupational Health Risks of Safai Karmacharies (Scavengers) Handling Solid Waste in Ghaziabad town (Uttar Pradesh).
- Recycling and reuse of organic solid waste through thermophilic aerobic method.

- Solid waste management for the domestic and commercial sectors.

Sanitation, Slum Improvement and other Social Issues :

- Consultancy on Low Cost Sanitation under Environmental Sanitation Programme in Chennai and adjacent urban areas in Tamil Nadu (India) (World Bank assisted project)
- Consultancy for community awareness and environment and health education (World Bank assisted project).
- World Bank assisted Kerala Urban Development Project: Trivandrum Sewerage and Sanitation Improvement Plan.
- World Bank assisted Low-Cost Sanitation Alternatives for unsewered areas of Madras city and adjoining village councils.
- Low-Cost Sanitation in Kanpur and Mirzapur towns in Uttar Pradesh.
- Evaluation of low-cost sanitation programme in Madhya Pradesh.
- Making 19 towns scavenging free in Assam.
- Low-cost sanitation to make Imphal (Manipur) scavenging free and training and rehabilitation of scavengers.
- Low-cost sanitation projects for Varanasi, Kanpur and Allahabad towns under Ganga Action Plan.
- Socio-economic impact on scavengers relieved from the task of manual handling of human waste in two towns in Punjab.

Water Supply:

- Uttar Pradesh Integrated Rural Water Supply and Environmental Sanitation Project for the World Bank

Assistance: Development of criteria for selection of villages for the project.

- Consultancy for the World Bank assisted Punjab Rural Water Supply Project for 120 villages.
- Consultancy for the preparation of detailed project report for water supply, sewerage and storm water drainage of a sector in Dwarka.

Sulabh International Academy of Environmental Sanitation proposes to further promote the objectives and functions undertaken by SIITRAT. It envisages that it would have a much larger mandate of acting as an international centre of holistic learning and education of all issues related to promotion of sanitation. The objective is to treat sanitation as a major field of education and study, revitalising training programmes involving highly qualified sanitation experts, engineers, health professionals, social scientists, ecologists and scholars from all other related fields. The Academy aims at bringing sanitation into sharp focus of academic activities and research nationally, regionally and globally. Towards this end the Academy is being planned to be set up on an ambitious scale in a 44 acre plot in Gurgaon and is being envisaged to be a deemed university.

Under the aegis of Sulabh International Academy of Environmental Sanitation, a number of HRD activities have been undertaken during the last couple of years. Regular training courses have been organised for hospital nurses, doctors and technical officers of Forest Research Institute, Dehradun, officers and trainees from Human Settlement Management Institute (HSMI) and SID, HUDCO and many other departments of the Central, State and Local Governments Personnel and for NGO representatives from all over the country. Visiting fellows and students from various universities from India as well as abroad (countries like Japan, France, USA, etc.) have visited the Sulabh campus.

An International Training Workshop on "Sanitation Technologies" has been sponsored by the UN Habitat under its capacity building programme on sanitation technologies in African countries (from 31 January '05 – 12th February'05). This training workshop was organised by Sulabh International Academy of Environmental Sanitation with the following objectives:

- To orient the participants in achieving the Millennium Goals set in Habitat Agenda which were discussed in the Johannesburg Summit on Sustainable Development;
- To enhance the Knowledge and Skills in Project Planning and Implementation of Sanitation Technologies through On-site Construction, Technical visits and Maintenance;
- To expose to various Sanitation Systems being implemented in the 3rd World including Low Cost Sanitation (different toilets), biogas, biofertilizer, Water Treatment, Duckweed Technologies, Solid Waste Management, Environment Protection;
- Role of the Schools as Promotional Centres and to create awareness among Students and the Community in Sanitation and Hygiene;
- Rehabilitation of weaker sections – Women and Children – through Education, Vocational Training and formation of Self-help Groups for sustaining their livelihood.

The participants from six African countries included Engineers, Architects and Administrators, etc representing NGOs, CBOs, Civil Society Institutions, Private Sector, Central, State and Local Governments.

Following are the core concerns of the academic programmes and activities of the Academy :

- Research and development activities on alternative sources of energy-biogas, solar energy, wind energy, and energy generated from human excreta, aquatic weeds, agricultural and municipal wastes.

- Sanitation and pollution control (with emphasis on low-cost technology)
- Improvements in the design of Sulabh Shauchalaya for greater efficiency, lower cost and wider applicability.
- Creation of facilities for application, testing and demonstration of the state of the art knowledge in low-cost sanitation, and various sources of energy and energy conservation devices.
- Organisation of training and orientation programmes on various aspects of environmental sanitation for professionals and the field functionaries.
- Collaboration with allied organisations to strengthen the research and training activities in the field of environmental sanitation.
- Setting up of training centre for providing integrated training in sanitation, water supply, primary health care, waste management, low-cost housing etc.
- Providing consultancy services in areas like water supply, drainage, sewerage, urban basic services.
- Documentation and dissemination of knowledge – organisation of seminars, workshops, publication of booklets and books on sanitation, waste management, health, energy and environment, preparation of audio-visual educative material for stimulating public interest and creating awareness on subjects related to environmental sanitation.

Sulabh Mahila Avam Bal Kalyan Sansthan

The Sulabh Mahila Avom Bal Kalyan Sansthan runs two training centres, viz. one at Mahavir Enclave and the other at Nanda Block, where six month training is imparted to downtrodden women and girls on tailoring and embroidery, jute craft, beautician's trade and computer basics. The Sansthan attempts to arrange bank loans for trainees who wish to set up their own business and helps others in securing jobs.

International Centre for Women and Child (ICWC)

The International Centre for Women and Child (ICWC), another wing of Sulabh, works for betterment of slum dwellers and regularly conducts the following activities:

- Baby shows for slum children
- Debate competitions in schools and colleges on issues relevant to slums
- Food Processing Training Centres have been set up by ICWC in 6 blocks in Mewat area in Gurgaon – Nuh, Taoru, Firozpur Zirka, Hathin, Nagina and Punhana. Three month certificate training courses are conducted and the food products made by the trained women are marketed by ICWC to provide them financial help. A total of 300 women have already received training in food processing. Apart from this, unemployed youth are trained in the following vocations to help them acquire jobs in the market: Plumbing, Electrical and motor winding, Gardening, and Security guard duties. The placement of trainees is also taken care of by the Centre and a total of 600 youths have already been trained.
- **SANYUKT** Vocational Training Centre: this centre aims to educate and link young girls and women with income generating activities. Women empowerment through training in Computer education, Cutting & Tailoring, Soft toy making, Mehendi application, Paper bag making, Diya & Candle making is done to make the slum women self-reliant and economically independent. Free education is provided to slum children in the afternoons.

- Self Defence Programme for Women – In collaboration with the Delhi Police – Crime against Women Cell, training for self defence was also given to 93 women from slums of Delhi.

Sulabh International Institute of Health and Hygiene

The Sulabh International Institute of Health and Hygiene unit was established in April 1994. Since then, various development programmes have been implemented related to sanitation, hygiene, health and safe drinking water. More than 12,000 women volunteers from 240 urban slums in Delhi and other States have been educated and trained in the areas related to health and hygiene. One thousand women volunteers from urban slums of Delhi were trained exclusively on STD and AIDS. The Unit is also one of the select NGOs in New Delhi participating in the Pulse Polio Campaign.

The awareness and training programmes were regularly supplemented by publication of write-ups and books for urban slums for better comprehension of the subject matter by the common people who understand local languages better than English/ Hindi. These books have been translated into 18 regional languages recognised by the Constitution of India.

Toilet complexes have been developed as Happy Homes, where slum dwellers are provided free medical services along with facilities like bathing, toilet, urinal, washing clothes etc. Condom distribution and contraceptive tablets are also provided free of cost in these Happy Homes.

The unit is operating Health Centres, run at two Sulabh Toilet Complexes in Delhi, viz. one at Mahavir Enclave and the other at Prayog Vihar, where free medical services are provided to poor slum dwellers. Laboratory testing facilities, where pathological tests can be done are also available. A FAMILY MEDICAL FOLDER incorporating the yearly medical record of all the family members living in the

Jhuljhuli village in Najafgarh district and Prayog Vihar slum in Delhi is maintained and updated regularly.

Sulabh also propagates the SULABH TOTAL HEALTH CARE CONCEPT (STHC) which is an innovative approach, recommending that Healthcare services have to be integrated with programmes on Sanitation, Hygiene and Safe drinking water to have the desired reduction in the morbidity and mortality rates. Sulabh Institute of Health and Hygiene has contributed significantly in promoting sanitation in hospitals and developed a manual of bio-medical waste disposal.

SulabhENVIS Centre

The SulabhENVIS Centre is the Environmental Information System wing of Sulabh, which has been set up under the EMCBTA (Environment Management Capacity Building Technical Assistance) programme by the Ministry of Environment & Forests, Government of India. On 20th March 2003 the SulabhENVIS Node started functioning at the Sulabh International Institute of Health & Hygiene, and was later upgraded to an ENVIS Centre by the Ministry in October 2004.

The mission of ENVIS is to provide access to and enhance the use of environment related information worldwide. The mission of the ENVIS Centre at Sulabh International is to gather and disseminate Environmental information related to the select topics of hygiene, sanitation, environment, and sewage treatment systems and technology for enhancing, supporting, facilitating and bringing under one umbrella the huge databank on the aforementioned subject areas for the benefit of a diverse group of people, ranging from students, policy makers, researchers and academicians to industries and institutions at national and international level.

The on-going programme of ENVIS, aims at providing a comprehensive environmental information data base under the identified subject areas, local conditions and issues pertaining to the environment of the country through a web-based database management system and a quarterly newsletter that is circulated to concerned organisations.

The Centre is actively involved in:

- Developing and updating an online database and supporting a query response system on the topics assigned by MoEF.
- Website www.sulabhenvis.nic.in – launched in May 2003. SulabhENVIS Centre was one of the first to upload its website on the NIC Server of the Government of India.
- Printing and distribution of Newsletter to different ENVIS Nodes & Centres, National & International Organisations and maintaining an E-newsletter on the website.
- Distribution of Posters, Pamphlets, and manuals on topics related to health & environment in Urban Slums.
- A survey of hygiene and sanitation status in slums of Delhi is being done.
- Analysis of the impact of Sanitation of Human Development Index is being done by the Centre.

Encyclopaedia of Sanitation

Sulabh has compiled an Encyclopaedia of Sanitation which is a valuable data-base containing information gathered from various countries pertaining to sanitation practices followed there and the history of sanitation.

Museum of Toilets

Museums, as repositories for the preservation and exhibition of the objects of historical, scientific and cultural interest, are found all over the World. But rare are the museums that display the evolution of toilets and their various designs. Until recently there was one small museum of toilets in Austria. The idea of setting up museum of toilets in the sprawling Campus of its central office at Mahavir Enclave, Palam Dabri Road in New Delhi, had been engaging Sulabh's attention for long, eventually leading the organisation to make hectic worldwide search for minutest details of the evolution of toilets, as also of various toilet designs used in different countries at different points of time. More than 100 Embassies and High Commissions of different countries based in New Delhi were contacted. Letters were sent out to each of them with a request to furnish information on the subject as also to provide details / photographs of various toilets designs used in their respective countries. More than 60 Embassies and High Commissions responded to the request and sent valuable information. Sulabh received some encouraging letters from various countries including USA and UK. The Austrian Embassy in New Delhi, played an extremely significant role in helping to materialize Sulabh's intention to setup the museum and sent some valuable information. Several other embassies contributed pictures as well as relevant literature. On its own, Sulabh made intensive worldwide search for the materials (artifacts, literature and photographs etc.) of historical significance. The intense search resulted in the accumulation of a significant reservoir of information on the subject. It nearly took over three years to fulfill Sulabh's ambition to establish the museum.

The Museum has been established with the following objectives :

- To educate students about the historical trends in the development of toilets;
- To provide information to researchers about the design, materials and technologies adopted in the past and those in use in the contemporary world;
- To help policy makers understand the efforts made by their predecessors in this field throughout the world;

- To help the manufactures of toilet equipment and accessories in improving their product by functioning as technological storehouse; and
- To help sanitation experts learn from the past and solve problems in the sanitation sector.

The Museum has a rare collection of factual information, pictures and objects detailing the historic evolution of toilets from 2500 BC to date. It provides a chronological account of developments relating to technology, toilet related social customs, toilet etiquettes, prevailing sanitary conditions and legislative efforts of the times. It has an extensive display of privies, chamber pots, toilet furniture, bidets and water closets in use from 1145 AD to the modern times. It also has a rare collection of beautiful poems related to toilet or its usage or call of nature.

The pictures displayed at the Museum make one aware of how the world looked like when societies did not have the benefit of water closets (W.C.) and the changes that have been brought about by its invention. Ornately carved and painted urinals and commodes attract attention and are a source of amusement to many. The pictures of medieval commodes are noteworthy. There is also a picture of a medieval mobile commode in the shape of a treasure chest, which the English used while camping out for a hunt. One could imagine the shock registered by some unsuspecting highway robbers, if they made away with such "treasure chests", thinking it to be containing something absolutely unprecedented inside. The Museum also displays how the Roman emperors used to have toilet pot made of gold and silver. The Museum has a rare record of the flush pot devised in 1956 by Sir John Harington, a courtier during the reign of Queen Elizabeth I. The Museum displays sewerage system of Mohenjo-Daro and Harappan civilization and maintains a detailed record of how modern toilet pans have emerged over a period of time.

The Museum has a stock of interesting anecdotes associated with the development of toilets. Tracing the history of toilets from Indus Valley Civilisation in Lothal, 62 kms from the city of Ahmadabad, India where a highly developed drainage system existed, the Museum documents the facts relating to some countries in Europe where most of the early technological developments in the evolution of toilets took place. The national flags of different countries, from where the pictures of toilets have been collected, are also displayed.

The Museum receives a daily stream of visitors from India and abroad. Most of them have found the project inspiring and unique. Sulabh is busy collecting more information from all over the World. The aim is to develop the Museum of an international standard.

Sulabh's March Towards Rural Sanitation

Sulabh International is one of the premier voluntary, non-governmental organizations engaged in the area of rural sanitation. Sulabh has already designed several models of rural latrines suited to different socio-cultural, physical, geological and hydro geological conditions of villages in different parts of the country. Sulabh has been advocating the promotion of two-pit pour flush toilets (popularly known as Sulabh Shauchalayas) designed for rural areas. The cost of these toilets ranges from Rs.600 to Rs. 3000 and the beneficiaries have the option to choose a particular design according to their economic conditions.

The Sulabh type individual toilets are an ideal solution to open air defecation in rural areas, where problems of insanitation abound, and where excreta disposal is a matter of letting nature take its own course. Sulabh has launched a sanitation promotion programme for sanitary latrines in rural areas. Although the technology needs no change, yet the programme has to be tailored to suit the rural requirements. Alteration in the design and use of local materials is required for reducing the cost. Sulabh has prepared several alternative designs both for superstructure as well for the leach pits for disposal of the waste with a view to satisfying people of different economic levels as well as to suit their socio-cultural habits. These have been designed and constructed giving the people a wide range of choice. Sulabh has already initiated a well planned programme for popularization of Sulabh-type individual toilets in rural areas. The important components of this programme are motivation, communication support, community involvement, health education and publicity through audio-visual aids.

Sulabh has taken up programme of rural sanitation in 436 districts of the country, where volunteers have been trained in the technology, methodology, implementation and follow up work

Presently, it is engaged in the implementation of rural sanitation programme in the States of Bihar, Orissa, Haryana and Uttar Pradesh. It is also engaged in a country-wide promotional campaign involving dissemination of information, motivation of potential users/adopters and extension education and follow up in other States too. A campaign has been launched to make the people in rural areas aware about the need and significance of personal, household and community sanitation. The organisation has published sanitation literature in 24 languages which is being distributed by Sulabh volunteers in villages all over the country. The beneficiaries of the low cost rural sanitation programme under the Centrally Sponsored Rural Sanitation Programme (CRSP) are helped by Sulabh volunteers in obtaining subsidy for the construction of household latrines.

In the States, where rural sanitation programme is in operation Sulabh International takes the entire responsibility for collecting the application, getting it processed and undertaking the construction of Shauchalayas. At the time of construction, the house owner is requested to keep a watch on the quality of construction and material used by

Sulabh, and after the work is finished, the house owner is required to fill a form confirming the date of completion of the work, stating whether he is satisfied with the work or not.

After completion of the work, Sulabh issues a guarantee card to every householder. This ensures free rectification of construction defects or deficiencies, if any, noticed during the next five years. Sulabh thus ensures trouble-free functioning of the Sulabh Shauchalaya. The house-owner is also educated about the maintenance of the toilet. A sample checking of completed toilets is undertaken periodically by Sulabh volunteers as a follow up. Sulabh has built a special infrastructure to ensure quality construction and render satisfactory follow-up services.

A pioneering project in collaboration with M/s Modi Industries was launched in 1991 to ensure diffusion of Sulabh Shauchalayas in individual households in Gurgaon District in the State of Haryana. Five villages of the Sohna Block were selected as model villages for construction of such latrines in the hope that the households adopting the technology would act as reference models for emulation by inhabitants of other villages. The funds for this project were made available partially by M/s. Modi Industries and partially by the individual house-owners. This has been the first project of its kind undertaken for a cluster of villages with assistance from an industrial house. Sulabh served as an action catalyst in the implementation of this programme. The organization provided a supportive environment by creating a felt need for this technology and motivating the villagers to accept and utilize the services provided. The programme was executed by properly trained, technically equipped and socially dedicated workers.

In recent years Sulabh has taken up various programmes such as health check-up camps, health and hygiene education campaigns in several villages of Maharashtra,

Uttaranchal and Haryana. Sulabh has implemented a project on "Sustainable Development and Health Environment" in villages of Uttar Pradesh, Haryana, Punjab and Rajasthan. This has been sponsored by the WHO and its implementation facilitated by the Ministry of Rural Development, Government of India with the main objective of promoting safe drinking water, sanitation and rural water supply.

Immediate acceptability of this low-cost appropriate technology among a large section of people was found to be attributable to the effective role played by Sulabh International. The organization proved capable of ensuring the diffusion of this innovation. The entire programme was conceived in consonance with the felt needs of the people and with their active participation. The adoption was neither random nor transitional. In a way it manifested a bottom up revolution in the field of rural sanitation.

As far as community latrines in rural areas are concerned, Sulabh International has its own reservations. Because of high-costs involved in construction and maintenance, it finds them economically unviable and unsuitable to the socio-cultural ethos in the countryside. Sulabh's reservations are based on its wide experience of constructing and maintaining thousands of community toilets in cities throughout the country. Sulabh has found that community latrines constructed by municipal authorities are not properly maintained, and hence not generally used, even though the local authorities possess organized infrastructure. In rural areas, panchayats do not have this infrastructure and the water needed for the use and upkeep of community latrines may also not be available. Sulabh's opinion has been endorsed by the Technology Advisory Group (TAG India) of the World Bank which carried out the feasibility study of low-cost sanitation in 3,600 villages in 13 States in the country in the years 1984-86. The study concluded that by and large the community latrines in rural areas were not being used due to improper maintenance and non-availability of water.

The Task Force of the Planning Commission, while outlining the approach to rural sanitation in Eighth Five Year Plan recommended that community latrines should not be constructed in rural areas. In view of the above, the Sulabh International advises that unless proper operation and maintenance is ensured, community toilets should not be provided in rural areas.

However, Sulabh is of the view that for demonstration purposes two to four seated community toilets may be constructed in schools, anganwadis, health sub-centres, panchayat bhawans and community development centres etc. in order to let people know about the importance of sanitary latrine and its benefits, eventually leading the people to get motivated to adopt the system in their houses. Sulabh favours the construction of community latrines at Block Development Office, market yards and bus depots in rural areas as there is likelihood of collection of adequate amount for proper maintenance. Sulabh is also advocating for the provision of two-pit pourflush toilets along with Indira Awas Yojana or other housing schemes in the rural areas.

For the speedy implementation of sanitation programme in rural areas Sulabh has made the following proposals to the Government:

1. A nation-wide campaign is necessary to make people aware of the adverse effects of open air defecation which is responsible for infections and a number of diseases. The people should be motivated to put on shoes/chappals while going for open defecation as this will save them from parasitic diseases. This type of education will be necessary until toilet facilities are available in the villages.
2. People in rural areas going for open defecation should be educated to put soil on human excreta after defecation so that flies may not sit on nightsoil and become carrier of diseases like diarrhoea,

dysentery, dehydration and cholera etc. which is responsible for death of about half a million children every year in the country.

3. Experience shows that it is not practicable to impose uniform design on users. There should be several designs of pour flush toilets meant for families below poverty line or those belonging to middle income group or higher income groups. Sulabh has prepared 46 designs which form part of the Project Report. Choice of designs should be left to beneficiaries who will take a decision, keeping in view their resources.
4. The possible help either in the form of subsidy or loan or both should be extended to all beneficiaries who want to have toilets in their homes irrespective of income groups. The first-come-and-served should be the criterion for the selection.
5. Programme of construction of toilets in rural areas is linked to making people aware of sanitation standards and health impact of insanitary conditions which require house-to- house contacts and follow-up. This can be effectively done by the NGOs. The role of the NGOs is very crucial in the implementation of the sanitation programme throughout the country.
6. The NGOs should be identified either by the State Governments or the District Administration. The selection of the NGOs should be based on their experience, expertise and infrastructure.
7. The NGOs identified for implementation of the programme should be given proper training in various aspects of the programme. A comprehensive programme of giving them training at various levels has been prepared which forms part of the Project Report.
8. The entire range of training starting with information, education, communication,

implementation and follow-up should be given to the same NGO. It has been experienced that if the work is divided among various organisations, it becomes a case of divided responsibilities that hamper progress of the work.

9. The NGOs selected for sanitation work should be allowed 15% of the estimated cost of IEC activities, including implementation and follow up services. In addition, 10% of the project cost should be allowed for training and support services, and publicity which would include printing of booklets, literature, posters, organising dramas and nukkad show etc. For setting up sanitary marts and production centres, subsidy as admissible under Government of India's Sanitation Programme should also be paid to the NGOs.
10. Interest-free bank loans should be provided to all beneficiaries who want to have toilets in their houses.
11. A nation-wide campaign should be launched through media for generating demand for safe hygienic toilets. The people would, thus, be saved from diseases which occur on account of open air defecation and insanitary toilets. The women will also be saved from the agony and humiliation of open air defecation. Apart from door-to-door campaign, it will be the responsibility of the NGOs to publish literature in local languages and make them available to the people. For wide publicity of the sanitation programme among the beneficiaries through pamphlets, booklets, calendars etc. 10% should be given to the same NGO which will be identified for implementation of the sanitation programme.
12. The implementation of the programme will require social mobilisation on a large scale which will include people belonging to various groups. Politicians and policy makers will have to be involved in a big way so

that they can take interest in policy decisions in favour of sanitation programme. The social mobilisation would also include village elders, doctors, lawyers and school teachers. Print media, radio and television will have to be fully involved in the implementation of the programme.

13. School sanitation should form an important component of the programme as it is through school teachers and students that the message of sanitary toilets will reach the villages. The NGOs which have been identified by the State Governments or district administration, in a particular area for construction of toilets should also be allowed to do construction and maintenance work of toilets in schools in that area, as also to impart training about maintenance and construction to teachers and students.

14. Public toilets in rural areas are generally not favoured. Public toilets, however, should be constructed near Panchayat bhawans, village markets, health centers, bus-stands and other public places where people congregate in large numbers. Such public toilets shall be constructed by the NGOs and maintained by them on the 'pay & use basis'.

15. The following recommendations of the Planning Commission in the 10th five Year Plan should be accepted in implementation of the rural sanitation programme successfully.

    a) Under the existing arrangement subsidy is available only for substructure. The Planning Commission has re-commended that the subsidy should be inclusive of sub and super structure for the basic twin-pit pour-flush system.

    b) The subsidy for low-cost household toilets should be given to rural below poverty line families and it should be at par with subsidy in the urban households. The Planning Commission has recommended that the present subsidy level should be increased to 50%.

Landmark Years in the Sulabh Sanitation Movement

1968 The author joined the Scavengers' liberation cell of Bihar Gandhi Centenary Celebration Committee. He established low-cost sanitation technology of two-pit-pour-flush toilet which is affordable, appropriate, safe, hygienic, culturally acceptable and better alternative to scavenging in India.

1970 A non-profit voluntary social service organisation, named as Sulabh Shauchalaya Sansthan, now known as Sulabh International Social Service Organisation was founded.

1973 Gradual evolution of the Organisation as catalytic agent for sanitation and social change. The organisation worked as catalytic institution between Government, local bodies and beneficiaries.

1973 One giant step towards scavengers' liberation. The first Sulabh Shauchalaya was established for demonstration in Arrah town (Bihar), which alone provided a comprehensive solution to end the inhuman practice of carrying as head load the excreta of another.

1974 A Sulabh public toilet on "pay-and-use" system was introduced by Sulabh in Patna for financial self sustainability of the toilet complexes.

1978 A National Seminar on the Abolishing of Scavenging was organised with joint collaboration of Ministry of Works & Housing, Government of India, WHO and UNICEF.

1979 Biogas from human excreta. The Organization was the first to generate biogas from public toilet and utilized it for cooking, warming bodies, lighting lamps etc. The biogas converted into electricity had lit the entire area of Gandhi Maidan, Patna, and Bus Stand in Ranchi, Bihar.

1980-81 As per implementation of Protection of Civil Rights (PCR) Act 1980-81 whole town approach to end scavenging was introduced in two towns of Bihar (Biharsharif and Purnia) for the first time in India.

1984 Sulabh International Institute of Technical Research and Training came into existence to give technological support to the Sulabh Sanitation Movement.

1986 Training and Rehabilitation programmes for children of scavengers in different trades (shorthand, typing, motor driving, mechanic, masonry work, carpentry, cane work etc.) were introduced to enable them gain employment. The author had persuaded the Union Ministry of Welfare through the Government of Bihar to extend the facilities like stipends etc. to the Muslim and Christian scavengers also for training and rehabilitation. Earlier these were not available for them.

1988 The author successfully led Harijans to Nathdwara temple in Udaipur, Rajasthan- His courageous act to defy inhuman orthodoxy and promote social integration was widely appreciated.

1989 Common *puja* and community meals together of high caste Brahmins and scavengers were organised by the author.

1990 Sulabh came up with an innovative approach to socially upgrade the scavengers' status in society. About 10,000 scavenger families have been socially adopted by high status people to interact with them as family friends.

1992 Launching of an awareness campaign. In February '92, the author organised a "National Seminar on Liberation and Rehabilitation of Scavengers"

- Priority was given to providing quality education and vocational training to children of scavengers.
- English-medium School was opened in Delhi for sons and daughters of scavengers. This is the first school of its kind in the country.
- Sulabh started work on a project to provide 1000 scavenger families with land from government on nominal price and loan from government agency.

1993 Sulabh Model of human excreta based biogas plant was approved by the Ministry of Non-Conventional Energy Sources, Govt. of India, for implementation through its state nodal agencies.

1994 Sulabh moved over to villages adopting Shahbad village in New Delhi with 200 scavenger families to provide integrated socio-economic development. Village Teekli (Haryana) was also adopted.

- Sulabh International Institute of Rural Development, Research and Training was started. It has the objective of utilising science for the service of humanity by taking technologies to the doorsteps of common man in India. The organisation developed multifaceted programme of training village youth (boys and girls) by which two youths from a village are trained in several vocations (social forestry, toilets, handpump, health care, carpentry etc.) so that they have capacity to earn their livelihood.
- Sulabh International Institute of Health & Hygiene was started to develop integrated healthcare and sanitation model of urban slums and villages. Concept of "Healthcare centre in Sulabh toilet complex" started at Harinagar (New Delhi) and a Slum Prayog Vihar (New Delhi), which was converted into a "Model Developed Slum" with all health facilities (immunisation,

family planning, antenatal care, health check up etc.)

1995 An innovative programme to train women on "Health, Hygiene & Sanitation" was launched. Within six months, five batches were trained in "sanitation, first aid & healthcare" in collaboration with USAID. About 12,000 women were trained to educate other women of urban slums of Delhi on environmental sanitation and community health.

- Publication of a Manual on *'Women & Sanitation"* in joint collaboration with USAID during training programme of women volunteers.

1996 The United Nations Centre for Human Settlements (UNCHS) at its Habitat-II conference, Istanbul, Turkey recognised "cost effective and Appropriate Sanitation Systems - Sulabh International Project," as a Global Urban Best Practice.

- Cleanliness and Sanitation Project for Railway Stations across the country under the Ministry of Railways, Govt. of India; Airport Authority of India, and Deen Dayal Upadhyaya Hospital, Delhi, was entrusted to Sulabh International Social Service Organisation.
- Sulabh International was granted Special Consultative Status by the Economic and Social Council of United Nations.
- "Action Sociology" was introduced in some Indian Universities, at the author's initiative.
- Jhuljhuli in Nazafgarh (N. Delhi) was developed as a model village. Mayors from USA visited this village and opined that two things impressed them most in India, one was Taj Mahal and the other was village Jhuljhuli.
- With Family Health International / AIDSCAP, Sulabh conducted programmes for creating a

cadre of 1000 women volunteers of urban slums to undertake S.T.D. and HIV/AIDS awareness campaigns in 40 urban slums.

- The Water Supply Sanitation Collaborative Council, Switzerland, selected Sulabh as its nodal agency for implementation of its programmes in India.

1997 Sulabh developed a new technology for composting biodegradable wastes. The technology known as Sulabh Thermophilic Aerobic Composting (STAC) requires only 8-10 days to degrade any biodegradable material without churning.

1998 Sulabh set up a People's Commission on the Abolition of Scavenging. Prime Minister Mr. Atal Bihari Vajpayee inaugurated the functioning of this commission at a function organised on November 26, 1998. Justice M.N. Venkatachalliah, former Chief Justice of India accepted the Chairmanship of the Advisory Board of the Commission.

1999 Sulabh developed a technology – Duckweed based waste water treatment and pisciculture, in collaboration with the Central Pollution Control Board, New Delhi, for the treatment of sewage. Later the CPCB made guidelines for use of the technology in different states.

2000 The programme of Sulabh Towards Villages, a national campaign to improve rural sanitation was launched.

2001 A country wide programme for involvement of women in sanitation, health and hygiene was started.

2002 Sulabh developed a new and convenient technology to make biogas plant effluent free of colour, odour and pathogen. The technology known as SET (Sulabh Effluent Treatment) lowers down BOD (Biochemical Oxygen Demand) of effluent to less than 1 mg/l, making it suitable for agriculture, aquaculture purposes or for cleaning floors of public toilets or for safe discharge into river or any water body.

2003 Steps have been taken to establish Sulabh Academy of Environmental Sanitation. Preparation for the publication of an Encyclopedia on Sanitation started. WASH Campaign started in collaboration with Water Supply and Sanitation Collaborative Council, Switzerland, and Sulabh Vocational Training Centre for the liberated scavengers was set up at Alwar, Rajasthan.

2004 Sulabh started working in Afghanistan and initiated the process of undertaking sanitation work in Nepal, Algeria, Uganda, Kenya, Ethiopia, Tanzania, Sri Lanka, Laos and Vietnam.

2005 The Sulabh International Institute of Technical Research and Training (SIITRAT) was transformed into Sulabh International Academy of Environmental Sanitation to strengthen and expand Sulabh's endeavours on scientific and technological research and training in the field of environmental sanitation.

Appreciation and Recognition

In recognition of its pioneering contribution to low-cost sanitation technology, liberation of scavengers, prevention of environmental pollution, creation of non-conventional sources of energy, training of change agents for voluntary action and diffusion of innovations for promoting inexpensive and affordable indigenous technology in sanitation, health, hygiene and rural development, Sulabh International has received enviable amount of national and international appreciation and applause. Its efforts have been acclaimed by State Governments, Central Government and various national and international agencies like WHO, UNICEF, UNDP, World Bank, UNCHS and HABITAT etc. The Governments of South Africa, Sri Lanka, Nepal and

Bhutan have invited the organisation to assist them in the implementation of sanitation and environmental improvement programmes in their countries.

The national and international media have widely reported the achievements of Sulabh International. The Washington Post, The New York Times, World News, The Nation (Bangkok), The Globe and Mail (Canada), Stuttgart Zeitung (Germany) and L'Osservatore Romans (Italy) have published write ups praising Sulabh's achievements. There is hardly any Indian newspaper and magazine which has not reported the programmes and activities of the organisation. The electronic media including BBC (London), Doordarshan and All India Radio have reported the achievements and contributions made by Sulabh International. The Doordarshan has made a documentary on the works carried out by the organisation. It often telecasts the programmes of the organisation to educate the masses on environmental sanitation and eradication of scavenging. The achievements of the organisation have been recorded in several books and professional journals like India File by Trevour Fishloe, Basic Needs of the Urban Poor by P.J. Richards, The Provision of Communal Service, Biography International, Reference Asia, Water World etc.

The United Nations Centre for Human Settlement has recognised Sulabh's cost-effective and appropriate sanitation system as a 'Global Urban Best Practice' at the Habitat-II Conference held at Istanbul, Turkey in June 1996. UNCHS (Habitat) has also conferred the 2000 Dubai International Award on Sulabh for the "cost-effective and appropriate sanitation system" for improving the environment. The Economic and Social Council of United Nations has granted Special Consultative Status to Sulabh in recognition of its outstanding service to people, specially those living on the edge-socially deprived, unempowered and disenfranchised. Observing the success of Sulabh's mission of environmental sanitation, the organisation has been chosen to be the nodal agency for the India Chapter of the Water Supply and Sanitation Collaborative Council (WSSCC) for the sanitation sector.

Affordable Sanitation in India

Much defecation in India still occurs in open spaces. But pioneering work by Sulabh International, a non-governmental organisation (NGO), has shown that human waste can be disposed of affordably and in a socially acceptable way. Sulabh's approach is based on partnerships with local governments, backed by community participation, and has substantially improved environmental quality in rural and urban slums inhabited by poor people.

Sulabh's solution is a low-cost, pour-flush water-seal toilet with leach pits for onsite disposal of human waste. The technology is affordable for poor people because designs suit different income levels. Flushing requires only 2 litres of water, compared with the 10 used by other toilets. Moreover, the system is never out of commission because there are two pits–so one can always be used while the other is being cleaned. The latrine can be built with locally available materials and is easy to maintain. It also has high potential for upgrading because it can easily be connected to a sewer system when one is introduced in the area.

Since 1970 more than 1 million of the units have been constructed in houses. In addition, 5,500 have been installed in pay-and-use public toilets, staffed by an attendant around the clock who supplies soap for washing hands. The public toilets include facilities for bathing and doing laundry and offer free services to children and disabled and poor people. As a result more than 10 million people have received improved low-cost sanitation, and 50,000 jobs have been created.

Sulabh's door-to-door campaigns also provide free health education to millions of people. The organisation trains local people to construct more latrines themselves, and has helped set up and maintain fee-based community toilets.

The UNDP, in its Human Development Report (2003) has recommended that Sulabh latrines must be adopted by international agencies as a model for widespread promotion of environmental sanitation in developing countries. While commenting an affordable sanitation in India, it published a box item as below :

Source : Human Development Report 2003 p. 105

Sulabh Sanitation Movement – A Low cost Solution to Success

Nearly 80 per cent of the country's population still either defecate in open or use unsanitary bucket latrines or smelly public toilets as per one estimate. This is true even in urban areas where hardly 20 per cent of the population has access to water/flush toilets connected to a sewerage system and only 14 per cent enjoy water-borne toilets connected to septic tanks or leach pits. In rural areas a mere 3 per cent of the population has access to sanitary toilets. This lack of adequate sanitation is responsible for several health problems. Cholera, dysentery, typhoid, para-typhoid, infectious hepatitis and many other diseases can be traced to the unsanitary disposal of human excreta. Lack of sanitation also has grave social consequences, the need to have 'night soil' removed has given rise to the profession of 'scavenging' or collecting it from bucket latrines, the streets and other locations. Though this practice has been banned and the Indian Constitution bans the segregation of those who service this profession, there are many pockets in the country where the practice continues unabated.

Sulabh International Social Service Organisation, a non-governmental organisation, founded by Dr. Bindeshwar Pathak, has in partnership with local Governments demonstrated the success of low cost sanitation technology throughout the country. Their solution called the *Sulabh Shauchalaya* is a low cost, pour flush, water-seal toilet with twin leach pits for onsite disposal of human waste. The technology has many advantages. It is affordable, even by the economically weaker sections of the society, and is designed to suit different levels of income. Flushing requires only two litres of water, instead of 10 litres needed by a conventional toilet. The toilet can never be out of commission, since, one of the two pits can always be used while the other is being serviced. The latrine can be built with locally available material. It can be conveniently upgraded as it is a stand-alone, on-site unit that can be connected to a sewer system as and when the latter is introduced in the area. So far, more than 700,000 units have been constructed or substituted for existing latrines in houses and more than 3000 have been installed as pay-and-use public toilets. The latter are staffed by full time attendant and provide facilities like soap for washing hands. 10 million people have been provided with improved, low cost sanitation and at the same time nearly 50,000 employment opportunities have been created in a commercially viable enterprise. As a social spin-off the enterprise has resulted in liberating about 50,000 scavengers from their enforced profession.

A key to the success of Sulabh Shauchalaya lies in creating public awareness and seeking community participation in implementing and maintaining the infrastructure. The organisation is also working with local groups on production of biogas from human excreta accessed from community toilets, and on generation of electricity. Its research and development activities are geared to seeking practical, low cost solution for solid and liquid waste disposal, including re-cycling in a financially sustainable manner.

Similarly, the National Human Development Report (2001) of the Planning Commission, Government of India made special mention of Sulabh Sanitation Movement :

Source: National Human Development Report, 2001, page 41

In recognition of the outstanding environmental achievements of Sulabh, the United Nations Environment Programme (UNEP), Nairobi, Kenya conferred "Global 500 Roll of Honour" in Beirut, Lebanon and UN-Habitat gave the "Habitat Scroll of Honour Award" for developing and instituting over one million twin-pit pour-flush toilets in India.

In view of Sulabh's appreciation, as above, it can be concluded that Sulabh Shauchalaya (twin-pit waterseal pour-flush toilet) has marked the beginning of an innovative low-cost sanitation technology in India. This technology has made an appreciable dent in India's problem of open-air defecation, where 750 million people out of the total population of over a billion, either defecate in open or use bucket privies or community toilets when available. The advent of this technology has provided relief to millions of people in both rural and urban areas, throughout the length and breadth of the country. Sulabh has made a new approach to the public sanitation problem not only by setting up a large number of community toilets and bathing facilities and converting bucket or dry latrines into two-pit pour-flush toilets, but also by initiating attitudinal change in respect of toilet behaviour and sanitation facilities. By sustained campaigns, education, and demonstration, Sulabh has been able to turn India's sanitation movement into a people's movement of health, hygiene and good living. The Sulabh Movement has also resulted in providing gainful employment to several thousands of people in the sanitation sector. This has, in turn, led to poverty alleviation and raising the quality of life of the poor in the country.

The remarkable aspect of Sulabh's low-cost sanitation technology is its truly indigenous character and the attempt to solve India's sanitation and other attendant social problems in strict conformity with socio-cultural ethos of the country. Sulabh's Technological breakthrough inheres Indianness and there lies its triumph. Another significant feature of the technology is its social dimension, entailing systematic and sustained efforts not only to liberate men, women and children of scavenger community from a demeaning occupation of manually cleaning and carrying human excreta, but also providing them training in economically rewarding occupations, imparting their children quality education, and upgrading their social status through persuading people of the well-to-do castes and classes to befriend them as equals. In all these social equalisation measures, Sulabh has gained tremendous success. Sulabh has been able to change the thought and attitude of the Indian people towards sanitation and towards those who are engaged in sanitation work. Sulabh Shauchalaya has now become a household name in India and scavengers have gained respectability.

Chapter - VIII

THE PERSPECTIVE ON PLANNING PROBLEMS

The complex character of India's socio-economic transformation during the post-independence era of planned and programmed development endeavours, can only be understood through an unbiased and objective appreciation of the intertwined process of continuity and change; which indeed characterizes the culture of contemporary Indian society. The continuance of some of the profoundly perplexing problems of the past, and the onset of tremendous progressive changes in the field of science and technology, as also in social and economic walks of life, explains the oriental duality and the dilemmas of Indian society. Whilst the country is legitimately proud of its culture and civilisation, and rightly reveres its old traditions, beliefs and value systems, an overwhelming number of people are unfortunately confronted with problems of survival and sustenance in their daily existence. Amongst the long list of social, economic and political problems, the population and poverty related problems of environmental sanitation are also a major source of considerable concern. The lack of basic healthcare and sanitation infrastructure is one of the biggest challenges that confronts the country even now.

The dimensions of India's sanitation scenario are truly scaring when viewed against the backdrop of shortfalls in the sanitation infrastructure. India still has 135 million people who have no access to basic healthcare facilities; 226 million who have no access to safe drinking water, and over 650 million people who lack basic sanitation facilities. Another set of statistics reveals that out of 1000 million people, more than 750 million either use dry latrines, manually cleaned by a section of people called 'scavengers' or, use their surrounding areas for open-air defecation. The number of bucket or dry latrines in the country is approximately 76.4 lakhs comprising 54 lakhs in the urban and 22.4 lakhs in the rural areas, and 30% of the urban households do not have household latrines. The large mass of people without toilets use any available space, road sides, parks, lanes, bye-lanes and railways tracks for defecation. Due to non-availability of household toilets women cannot defecate in privacy with dignity. They go to open areas either before sunrise or after sun set, which is not always safe for them. During the rainy season, women face immense difficulties at night when going out for open defecation. In flood affected areas they are sometimes subjected to indignities as they have no option but to defecate in the open alongside men. The practice of open defecation, besides being a cause of considerable embarrassment, is also a source of infection, leading to a large number of diseases. The most affected are children of whom half a million die every year because of diarrhoea and dehydration. The age-old unhealthy practice of open defecation and use of bucket privies are potential hazard to health and hygiene and expose the entire community to foul environment and pollute the rivers and other water courses.

A word about the prevalence and persistence of the practice of open-air defecation in India will be relevant here. While nobody now approves it, the practice of open defecation has, however, been given religious sanction in several sacred books. There are edicts in the old religious scriptures which direct people to defecate at a place little away from home – at places far away (at a distance of an arrow-shot) from one's habitat. In the older days people rarely thought of making provision for a restroom inside the house. In tropical climatic conditions open surroundings were considered healthy and good enough for the purpose.

In the pastoral past people were advised to dig a little hole in an open field, defecate in it and cover it with earth or some other stuff like tree leaves etc. People did it gladly and found it healthier and safer way of human waste disposal. Besides religious sanction, the practice of open air defecation was then socially and culturally acceptable.

Now of course, the old context is no longer relevant. Toilet is a necessity, indeed, an essential requirement of modern living and hence, the issue of toilet is a matter of considerable concern for the country.

Looking at the gigantic size and magnitude of the problem of environmental insanitation, and the predicaments of the people without toilet facilities – 10.76 crore people in rural areas and around 1.41 crore in urban areas – everyone in India agrees that the problem requires the nation's utmost attention. The people in the Government readily concur with this view, but find themselves quite often helpless in mobilizing massive financial resources that are needed to tackle the situation.

No Latrine
(No. of household 2001 Census)

Total	122,078,136	(63.6%)
Rural	107,967,200	(78.1%)
Urban	14,110,936	(26.3%)

While the best solution to the problem lies in the extensive use of the sewerage system, which no doubt is the best way of disposing of human waste, the real difficulty is that of the crippling constraint of financial resources. Neither the Government nor the local authorities, nor even the beneficiaries can afford to bear the capital or the operation and maintenance cost. Another constraint of no less serious nature is that of water shortage, because sewerage needs considerable quantity of water for flushing the toilets. Therefore, sewerage system is not financially feasible or cost-wise practicable.

Septic tank, the other option is also quite costly and beyond the reach of the common people. Furthermore, it entails many drawbacks and operational problem of desludging after one or two years, which again requires the services of scavengers.

Closely connected to the scaring sanitation scenario is a typical indigenous social problem of national concern, namely, the problem of scavenging – the problem of manual disposal of human excreta. Scavenging, as prevalent in India, is the system in which human excreta is cleaned and carried away manually. The people doing this job are called 'scavengers'. For centuries, scavengers have been subjected to all sorts of dehumanising and discriminatory practices wrapped and rooted in the customary practices of untouchability. The condition of scavengers inheres innumerable instances of continuing degradation and their disadvantaged position in Indian society, quite often manifesting itself in indignity, exploitation and abuse. The Father of the nation, Mahatma Gandhi, and the chief architect of India's Constitution, Dr. B.R. Ambedkar, were deeply concerned with the pitiable plight of the scavengers who kept the houses and the cities clean, and yet were made to suffer myriad indignities.

Mahatma Gandhi tried to bring the scavengers into the mainstream of society, but the task remained unfulfilled in his life-time. In the evening of his life, in acute anguish, he wrote that if he were to be reborn, he would like to be born in a scavenger's family so that he could redeem the scavengers' situation. In free India, the constitution-makers, in due deference to the unfulfilled dream of the Father of the Nation, declared dejure abolition of all forms of untouchability in pursuance of the constitutional commandments of equity, equality and social justice. This, however, did not help resolve the problems of social discrimination and disrespect affecting the scavengers' community, whom the Indian society continued to treat as untouchables amongst the untouchables. After Independence the Government of India formed many high level committees to improve the living conditions of the scavengers, but tangible results did not emerge and the plight of scavengers remained as lamentable as ever before.

In the given scenario the task of making the country free from scavenging entails enormous difficulties. It calls for good planning, efficient administration, adequate financing, and proper implementation with back-up technology and follow-up. It will be prudent to carefully consider all these aspects and plan the programme accordingly. Several committees and commissions were set up by the Government of India in the past to find a solution to this problem; but the progress has been negligible. The reasons for this dismal situation are many. Discussion on various aspects of the planning processes and problems will facilitate the development and implementation of pragmatic policies and programmes for environmental sanitation and eradication of manual scavenging.

As explained earlier, the planning process for improving the environmental sanitation in the country had started following the commencement of the planned policies and programmes of development. However, the pace of progress has been slow. For a variety of reasons, sanitation could not be accorded a high priority, and hence, the sanitation scenario continued to be grim. The enormity of the challenge remained a daunting problem. Whilst the targets fixed for the purpose were chased energetically, there remained a huge backlog of people without adequate sanitation, more notably due to the lack of household latrines, because of which the problem of open defecation continued. Therefore, the abolition of manual scavenging and the liberation of scavengers through the conversion of dry/bucket latrines into water-flush latrines by an affordable, acceptable and easily available system of sanitation technology to do away with the manual handling of human excreta remained a difficult goal. Regardless of the Government's concern about the plight of the scavengers, and despite the implementation of several programmes for their welfare and well-being the desired outcome remained unsatisfactory. The barriers to progress in the sanitation sector were as follows :

- The basic issues concerning sanitation promotion had not been handled properly despite years of rhetoric, good intentions and hard work
- There had been little evidence of political will on the part of the political parties to deal with the problem. Lack of political commitment needed to create an environment conducive to sanitation promotion was indeed responsible for poor sanitation status of the country.
- Promoting low-cost sanitation facilities and hygiene education was accorded little prestige and recognition.
- Agencies responsible for creating a supportive environment for sanitation have had ineffective policies and poor implementation at all levels.
- The institutional arrangements put in place often had fragmented responsibilities in the multiplicity of government agencies and departments.
- Excreta management and hygiene education only attracted a fraction of the resources needed to do the job. Despite the recognition of sanitation being as important for health as is water supply, it received far fewer resources.
- Critical issues of behaviour were frequently ignored or handled badly.

- There had been a marked neglect of consumer preferences. Too often the government tried to promote measures, which people did not want or could not afford, or both.
- Engineers and health care professionals who are responsible for promoting sanitation were often unaware of effective promotional techniques, and hence the adoption of social marketing and participatory approaches to sanitation were missing.

Given the magnitude and importance of the problem, the barriers to progress can be grouped into the following linked and overlapping categories.

Lack of Priority

Notwithstanding the emphasis by the political leaders, planners and administrators, the total efforts made by the governments have fallen short of what is really required to liberate the scavengers and assign priority to the programme in the governments' planning process. In fact, the absence of priority in the various plans, has been the single biggest factor for which there are still 1,27,872 scavengers engaged in the demeaning task of handling human excreta even after 62 years of Independence. The 20-Point programme formulated first in 1972, which was the guide for fixing priorities of planned programmes did not include sanitation as an item, nor was it included in the 20-Point Programme of 1982. In fact even when the late Prime Minister, Smt. Indira Gandhi, delivered her historic speech in parliament in March 1983, in regard to eradication of the scavenging system, the technology of twin-pit pour flush waterseal latrine had been well established and successfully tried for conversion of bucket latrines. It was the late Prime Minister Shri Rajiv Gandhi who recognised the availability of this technology and included the eradication of scavenging and rehabilitation of scavengers in other suitable professions in the revised 20-Point Programme in 1986. This was perhaps the single most vital step, which should have been taken by the Government of India to bring the problem of scavenging and the plight of the scavengers and their families to the fore. This would have required the conversion of dry/bucket latrines into water flush toilets and providing such toilets to those households, which have no latrines even though space is available for their construction. For such households, which have no latrines and where space for providing water flush toilet is not available, community toilets were necessary in order to improve sanitation and avoid open air defecation, thus dispensing with the services of scavengers to keep the cities free from such practices.

Selection of Technology

Despite the availability of affordable sanitation technology option, our policy planners went wrong in giving preponderant emphasis on sewerage and ignoring the possibility of introducing the available two-pit pour-flush technology in even class I cities where the problem of scavengers has already been relatively more pronounced. The programme chalked out did not speak about the ideal suitability of two-pit pour flush technology in large scale in other cities and towns where sewerage was inconceivable even to the planners on economic grounds. The result is that we are nowhere near the target even two decades after the end of the 'Sanitation Decade' (1980-1990), although the funds for urban areas have generally been spent almost entirely on sewerage, the benefits of which have hardly been derived by the city population. The most unfortunate part has been that in the cities where sewers have been laid, city residents have mostly refrained from taking house connections to sewers due to financial constraints and therefore the dry/bucket latrines and system of scavenging have been continuing side by side in all such cities. Public health engineers had always laid emphasis on sewerage and water flush latrines and had not studied the alternative low cost options.

If the planners had set their priorities right on liberation of scavengers from the cities and towns of the country they would have allotted the larger part of the limited sector allocation for low cost option for urban sanitation to conversion of bucket/dry privies to leach pit latrines and construction of those latrines in the households where there exist none. Thus, once the target group community was selected for conversion/ construction programme, the planner had to select the technology appropriate to the needs and resources of the community. This selection was to be based on a combination of economic, technical and social criteria. The issues arising out of them can be reduced to a single question: "what is the affordable, socially acceptable, easily available, technically feasible technology that the users would accept and be able to maintain and the local authority is institutionally capable of operating?" Therefore, the constraints which arise in the selection of appropriate technology are: (a) Acceptability; (b) Affordability; and (c) Availability.

The answer to all those questions would have indicated on site-sanitation i.e. twin-pit pour-flush water seal latrines popularly known as 'Sulabh Shauchalaya' which would have addressed the issue satisfactorily to fulfill the requirements. The concept of the Sulabh Shauchalaya has now been adopted on a large scale almost all over the country. Even though the technology and solution are known, it is always necessary to have a good technical backup to provide solutions to any difficulties and problems which may arise during the implementation of the programme. Monitoring and over-all technical control at state and central level will ensure success of the programme.

Lack of Political Will

There is little political will and administrative determination to deal with the problem. Politicians rarely lose their job because of the poor implementation of sanitation programmes. Political commitment needed to create an environment in which demand for sanitation can grow remained at best feeble. The lack of an effective political will thus, became a cause and an effect of the problem, and a formidable barrier to successful sanitation promotion.

Further, the promotion of low-cost sanitation facilities and hygiene education had never been prestigious. Among the professionals, many of the best and the brightest avoided working on affordable approaches to excreta management because of the low-status and less recognition for the work. For most of them low-cost solution to excreta management had little prestige compared to the conventional sewer systems used by the World's affluent populations.

Poor Policy at all Levels

Agencies responsible for creating a supportive environment for sanitation, in general, have had ineffective policies at all levels. These include too much attention to water supply at the expense of excreta management and hygiene education; a focus on short-run outputs (hardware) rather than long-term behavioural change. More fundamentally, a practical approach to the problem, upon which sound policy could be based, was often lacking.

Poor Institutional Framework

The institutional frameworks in place often had fragmented responsibilities in a multiplicity of government agencies and departments. They neglect the needs of the most vulnerable segments of the population and ignored the powerful role that NGOs and the private sector can play. Therefore, the government failed to promote sanitation and revamp the existing institutional frameworks needed for the change.

Inadequate and Poorly used Resources

Excreta management and hygiene education attracted only a fraction of the resources needed to do the job. Both of these have received far less resources. Even when resources were made available, they went into hardware, and not enough was done for community mobilisation and hygiene promotion.

Inappropriate Approaches

Even when the promotion of sanitation has been attempted, the approach taken has been wrong. Frequently, attempts were made to find universal solutions. These have failed to acknowledge the diversity of needs and the cultural, economic, and social contexts in which they occur. For example, although the expectations of urban population often differ from those living in rural settings, the technological options offered had been the same. Critical issues of behaviour were frequently ignored or handled badly. Short-term "fixes" had been generally favoured over long-term solutions, and no lessons were drawn from the earlier experience. This situation was further aggravated by a lack of awareness among engineers and government decision-makers on the performance characteristics of on-site excreta management systems. This lack of awareness has been, in large part, due to the focus of traditional engineering education on conventional sewerage systems. Rejection of an on-site excreta management approach is also often based on the belief that the available "hardware" for on-site management is technically inferior, less sophisticated, and has a managerial and administrative burden on households and government agencies alike. Sanitation also fails by being defined and applied too broadly or too narrowly within a specific environment. In some cases, for example, the scope of environmental protection and pollution control becomes so broad that the focus on basic household excreta management is lost.

Neglect of consumer Preferences

Too often, the stress has been laid on the promotion of what people do not want or cannot afford or both. Low-cost technologies have often been viewed by the consumers as low-status technologies. Others found appropriate by their promoters, are far beyond the financial reach of those in most need. Further, much hygiene promotion has been based on messages that ignore existing knowledge, belief, and experience. Put simply, most of those engaged in promoting sanitation simply do not hear the people they serve.

Ineffective Promotion and Low Public Awareness

Although people have opinions about excreta management, they are reluctant to talk about the management of excreta. Thus, selling the idea of improvements in sanitation is difficult. Engineers and health care professionals who are responsible for promoting sanitation are often unaware of effective promotional techniques and continue with such approaches that alienate the "target populations" by denying their voice, desires, and involvement in the process. Those who are charged with promoting sanitation are seldom prepared to do so in their education of others or in their professional practice. Adoption of social marketing and participatory approaches to sanitation is promising, but is still in its infancy; we have much to learn.

Cultural Taboos and Beliefs

In most cultures, the handling of excreta is considered as taboo, and viewed as a disgusting or a dangerous nuisance not to be discussed openly or seriously or both. No one wants to be associated with excreta; even those who reduce its offensive characteristics for others are stigmatised by association. Problems cannot be solved if people do not want to talk about cultural taboos and beliefs and do not want to be associated with their solution. In many contexts, including modern technological ones, taboos block the safe recovery of valuable agricultural resources from human wastes. The excreta taboo lies behind many of the barriers to progress in this area. To counter this, sanitation promotion and hygiene education should link the value of excreta (faeces and urine) with ecology. They should promote an understanding of the essential role it plays in the life cycle of plants and animals, as well as the damaging effect that it can have on health and environment when improperly handled, discharged or reused.

Pre-requisites of Planning for Sanitation Promotion

The foregoing discussion on the barriers to an effective planning for meeting the sanitation challenge in the country is a pointer to the fact that the planned and programmed policies of sanitation promotion have had such structural and functional deficiencies which undermined the effectiveness of strategies and approaches adopted for the purpose. The global discourses on sanitation promotion reiterate and reaffirm an approach that is not only human-centred but also ecologically sustainable. Accordingly, it has to be concerned with equity and sustainability of the environment and the health, welfare and well being of the people. Its goal is to devise and develop such sanitation measures and institutional mechanisms, policies and procedures that are people-friendly and socially, economically and ecologically prudent and pragmatic. The key issues that are critical to designing successful sanitation programmes are :

- Ensuring access to safe sanitation systems for all communities.
- Promoting such sanitation systems that are cost-effective, safe and culturally acceptable to the users.
- Effecting genuine community involvement in the planning and management of systems.
- Disseminating information required for decision-making to all segments of the user communities.

The unprecedented sanitation challenge requires that new strategies and methods to improve sanitation be applied to ensure equitable access for everyone. This requires :

- Adoption of new and innovative approaches
- Application of a mix of technologies and systems
- Conservation and protection of resources
- Incorporation of community values, perception and practices
- Restructuration of sanitation as an essential public service
- Development and promotion of a range of low-cost sanitation technologies applicable to different cultural, environmental and economic conditions of user communities.

Keeping in view the limited resources and several other national priorities of pressing nature, the planning process for the liberation of scavengers will involve the following:

1. Estimation of the total number of scavengers still engaged in the demeaning task of handling human excreta.
2. Estimation of the number of scavengers already liberated in different cities and towns due to conversion and/or construction of low cost latrines along with their present occupational pattern.
3. Assessment of the existing number of dry/bucket latrines on-town-by town basis.
4. Augmentation of the process of converting/ constructing low cost latrines as an important measure for the liberation of scavengers.
5. Enlargement of the programmes of training and rehabilitation of the liberated scavengers and their wards.
6. Estimation of resources required separately for both the tasks, i.e. liberating the scavengers by converting/ constructing low cost – latrines; and training and rehabilitating the scavengers and their wards.

7. Identification of the financial, technological, administrative and other socio-cultural constraints.
8. Finalisation of ways and means to ensure proper implementation of policies and programmes of low-cost sanitation and the training and rehabilitation of scavengers and their wards.
9. Fixation of investment priorities in the budget allocation for specific projects pertaining to environmental sanitation and eradication of manual scavenging.
10. Formation of macro-plans for the provisioning of resources for a smooth and uninterrupted implementation of conversion/construction programmes of low-cost latrines; training and rehabilitation programmes of liberated scavengers and their wards, creation of adequate job opportunities for the liberated and trained scavengers; and assigning responsibilities to "Scheduled Caste and Scheduled Tribe Welfare Corporations" for an early rehabilitation of the liberated scavengers and their wards.
11. Enlargement and strengthening of research and development activities for cost reduction, pollution prevention, and technological advancement in the field of environmental sanitation.

The eradication of scavenging and liberation of scavengers by providing households with water seal latrines decidedly merits overwhelming priority from all points of view. There are some factors which are crucial for success of the programme. It is imperative to revise a strategy and consider the following issues;

1. Vitalising political will and policy commitment backed by decision making apparatus at the Centre, State and Local level.

2. Providing technical support with advanced technologies suited to different areas and situations.
3. Conducting research in the field of materials for cost reduction and easy availability, pollution aspects and development of new concept of technology and design.
4. Studying the financial aspects, in terms of availability of funds and the capacity to bear cost and repayment of investment by the householders and the local body.
5. Studying the legal tools and other enforcement problems.
6. Orientation, education and extension of the programme for acceptability of the programme by all concerned.
7. Monitoring of programme to get feedback for effecting improvements on financial and technical control and solving problems as the programme proceeds.
8. Preparing operation and maintenance guidelines.
9. Launching a strong movement for social reforms in order to mould the attitude of the scavengers to leave their age old occupation of scavenging and move into occupations not connected with scavenging.
10. Providing adequate facilities for training the liberated scavengers and their wards in different trades and skills and their rehabilitation.
11. Training the managers, sociologists, technologists, supervisors, skilled and non-skilled workers for implementing the programme.
12. Planning of suitable institutional framework for project preparation, mass construction and conversion, and arranging materials.
13. Making use of the mass and electronic media for the awareness campaign of the programme.

The Holistic Strategy of Sanitation Promotion

A forward looking system of planning for sanitation promotion emphasizes the need for developing a rational and holistic strategy of sanitation promotion that is capable of stimulating the demand and providing means and mechanisms for meeting that demand. The foremost requirement in this connection is to first treat the subject of sanitation holistically and deal with it in a focused manner. In formulating strategies, the following are the essential requirements.

- Greater attention to reaching the 'hard-to reach' groups who are largely poor, less educated (if not illiterate), living in conditions of squalor and disease, and are more resistant to change.
- Increased emphasis on the development and promotion of appropriate and indigenous technologies for home hygiene and environmental sanitation, keeping in mind that appropriate and low-cost technologies offer the best prospects of increased coverage for poor people. In respect of appropriate technologies, the latest Human Development Report (2003) observes: "In sanitation, there is a pressing need to provide technologies that people want to use, because decisions about sanitation are made at the household level".
- Sustained efforts to create an environment in which demand for sanitation can grow, and which in turn, can strengthen political will. The issue of political will is a key to successful sanitation promotion.
- Building of strong partnerships with private providers. This is of utmost significance in view of the fact that local governments, municipal bodies, Panchayats and other public bodies in poor countries are so much strapped for cash that they cannot fund large investment in sanitation infrastructure. All over the world the privatisation of water and sanitation has

occurred through private players and NGOs with tremendous good results.

- Involving NGOs in increasing the role of civil society organisation in providing sanitation services. This is important in view of the abundant evidence in both developed and developing countries where NGOs have been successful in filling gaps left by the public systems. They have also been useful in articulating community concerns, especially for people to make public institutions perform better. It is appropriate to remember here that in water and sanitation, the interest of the user communities has been served better with NGO support. But important message here is that NGOs should be a complement to and not substitute for state activities.
- Fixing realistic goals and targets as benchmarks for assessing progress.
- Energizing activities and strengthening the public system of providing sanitation to provide full range of sanitation services through increased public spending.
- Enlarging the availability of public toilets with different designs, for different kinds of users (children, women, aged and handicapped) and for specific locations – parks, schools, railway/bus stations, airports, and market places, places where religious gathering and exhibitions take place.
- Augmenting financing and funding of public toilets by way of raising investment funds, like sanitation bonds on the lines of infrastructure bonds.
- Improving the management of public toilets by adopting good management practices.
- Creating a cadre of trained sanitation workers for sanitation and toilet related jobs.

- Promoting alternative human waste disposal systems like twin-pit pour-flush toilets with on site disposal of excreta.
- Formulating national policy on sanitation highlighting policy perspectives; need, resources and action plan; national priorities; implementation agencies and international funding cooperation.
- Mounting public education campaigns through the mobilisation of community and local institutions for creating awareness on matters concerning health, hygiene, household cleanliness and environmental sanitation.

Translating Challenges into Action

The challenges are complex but by no means insurmountable. In responding to them, a concrete plan of action containing a vision, achievable targets, implementation strategy, and monitoring mechanism needs to be prepared at the country level. This is essential because many of the developing countries do not have national plans to transform their wish list of principles, requirements and guidelines into concrete action. There are others who do have plans but not the means of implementation. This requires strong political commitment and support as much as an enabling environment and financial resources. Considerable attention needs to be focussed on removing the existing policy-oriented, fund-centred and technology-related institutional constraints and impediments. And more importantly various cooperative actions are needed on the part of the governments, business organisations, civil society associations, international organisations and other relevant stakeholders. The major requirements in this connection are :

- Promotion of political will at all levels to ensure that integrated approaches to sanitation services are mainstreamed into relevant sector policies and programmes.

- Mobilisation of resources at all levels for sanitation infrastructure and services development.
- Development of strategies and action plans aimed at reaching sanitation targets.
- Encouragement to the development and implementation of affordable and appropriate sanitation projects in rural and low-income urban areas where needs are greatest.
- Development of effective sanitation and hygiene practices through advocacy, training and capacity development in partnership with municipal corporations, civil society groups, NGOs and the private sector.
- Promotion of knowledge about innovative sanitation technologies that are cost-effective, efficient and user-friendly through partnership with national, regional and local public forums and institutions directly or indirectly concerned with sanitation issues.
- Bringing about behavioural changes in sanitation and hygiene practices.
- Identifying best practices and learning from the experiences arising out of implementation of projects and programmes related to provision of safe sanitation services.

Chapter - IX

CONSTRAINTS IN IMPLEMENTATION

Despite the past and present efforts, the achievements of the country in improving the sanitation scenario have not been significant, when viewed in the backdrop of the fact that there still are 6 lakh EWS of bucket latrines (urban and rural) in use and nearly 37% of the urban population have no household latrine facilities and 57% of the rural population have sanitary toilets in their houses. The technology of low-cost pour flush toilets with on site sanitation, which had been successfully introduced about 30 years back in the congested urban areas and about 40 years back in the rural areas, would have made a vast difference in the sanitation of the country had it received the full support of the Government and the people. What has prevented the adoption of this technology and what difficulties and constraints have been faced needs to be examined before any solution can be thought of. Several factors have contributed to making the task difficult, but perhaps the primary reason being that it is not a "felt need' for the users to change over from bucket/dry privies to water flushed system. This single factor could be taken care of at least partly through health education, motivation and community participation. Lack of political will (giving priority to the programme), the attitude of the authorities, budgetary constraints and pattern of subsidy, technical snags, behavioural attitude of scavengers and householders, legal provisions and their enforcement, training and methodology of operation, have all been responsible in some way or the other in slowing the progress of the conversion of the bucket/dry latrines programme and thus liberating the scavengers from the demeaning task.

Felt Need for Household Latrines

Both history and culture reveal that provision of the latrine as an inseparable part of an Indian house was ignored. Returning human excreta direct to the soil by using open field for defecation was a matter of habit and convenience before the advent of modern civilisation. Rapid urbanisation brought home the need for a house latrine. Even so the household privy has not found favour in many homes. It is common to find many houses in towns and cities without any latrine although most of them have a bed room and a living room. This is obviously because there is no felt need or perception of the importance of having a latrine within the house. While undoubtedly a household privy might not be feasible technically for lack of space, or for economic reasons, but in many cases there is lack of desire to accord priority to a latrine and consider that a house is really incomplete without it. Where there is a "felt need" for a latrine many features of the house would be sacrificed for its sake. Health education, community awareness and, above all, motivating the family for the need of a sanitary latrine in the house, for health, convenience and for safeguarding the modesty of women in particular have to be carried out preferably with the help of voluntary social organisations. This is perhaps one of the primary constraints in the programme.

Social and Economic Dimensions of the Problem

The success of any programme depends on its acceptance by an informed community, as also on the degree of social emancipation and economic betterment it confers on that community. This is true also for conversion of bucket latrines and liberation of the scavengers. The urban householder with a bucket latrine looks upon it as a necessary evil. Sociologically he is conditioned to prefer a water-seal privy, provided the cost of conversion is within his financial capacity.

Where the conservancy service is heavily subsidised by the municipality by paying each scavenger a monthly salary and he is also paid by the individual household served by the scavenger, the local body itself is making the bucket latrine financially more attractive. The fact that his salary per month is met from the general revenues of the local body contributed by the same houseowners as property tax and other levies is seldom realised. When the cost of conversion into a waterseal latrine is also subsidised, the economic disadvantage is removed and the programme in such cases has made some headway. Here again the repayment of loan charges on the portion not subsidised is to be such that there is not much disparity between what a householder pays per month for scavenging and what he would be required to pay for the repayment of the loan.

In the case of houses without latrines, the householder is handicapped by his economic limitations rather than by any social inertia. When the householder is deprived of an individual latrine, legal injunction against open defecation becomes ineffective and punitive action is not feasible. If the community latrines are the solution to such householders the urban poor would prefer the water-seal type, over the bucket type provided these are kept clean all the 24 hours.

Householder's Attitude

The most difficult part of the operation is influencing the householder who has not only to be persuaded to accept a change but must also be prepared to shoulder the cost, however small it may be. Even if loans and subsidies are offered, the various procedural formalities are too cumbersome and time-consuming to scare him away. While the standard water-closet with connection to sewer is far more expensive, there is considerable scepticism about the design and working of the cheaper hand-flush toilet system. And then there is yet another extremely difficult operation of arranging men and materials for construction of the toilet.

Another reason for reluctance of converting the dry latrine on the part of the householder is the low cost of maintaining and cleaning the dry latrines at some places. The municipal rates in most of the local authorities are very low for cleaning the family bucket privy, which discourage a change over. If the local authorities give a substantial rebate on the scavenging/drainage tax to the owner who switches over to a water-flush toilet, there will be a better response. Where service is municipalised, there is need for raising the charges for cleaning the service latrines substantially. In states where there are private arrangements, and scavenging charges for the latrines are high, it has been observed that there is more eagerness and response from the households for the conversion of the bucket privies. Amongst those who accept the idea of conversion of dry latrines, there are many who have seen the conventional water flush latrines and they insist on having a flushing cistern, the use of wider ceramic pans and conventional traps. They do not feel happy with hand flushing.

Setting of latrine and particularly the pits is a difficult and time consuming task. The convenience of the head of the family and meeting him at the time when he is available in the house, are the important pre-requisites. Even when the availability is ensured, it is often found that he is reluctant to agree to the construction of leach pits in the available vacant space. He prefers to use that piece of land for extension of his living accommodation rather than waste it for the construction of latrines or pits. He expresses many doubts such as (i) the building may get damaged by the water from leach pits; (ii) his or his neighbour's water source may get polluted by the leach pit; and (iii) in case they adopt this system, the sewers shall never be laid in the area. Many a time they say that some important politician or the minister himself had given an assurance that sewer scheme would soon be taken up and therefore they would not go for any other system which was inferior to sewerage. The need for special type of pan and trap, specific design of leach pits, its size and shape depending upon the site condition etc. have to be explained convincingly.

Experience has shown that the programme could run into trouble if the householders do not cooperate fully. Some times they do not sign the agreement form or on completion of the latrine unit do not readily give the certificate that they are satisfied with the construction. Thus the payment of the construction agency gets held up and the programme suffers a set back. At times due to improper use of the toilet, the outlet pipe gets choked and the householders pass on the blame to the construction agency for having executed faulty construction. The complaints are made to the local authorities without caring to inform the construction agency for rectification of the defect. The defects remain unattended. This is how the programme gets a bad name and tends to become unpopular, giving an impression that the system is faulty. All these points need to be clarified and suitably explained through education and motivation.

The local authorities deal only with owners of houses in sanctioning a subsidy and loan for conversion of dry latrines into pour flush units. The loan amount should be repayable by the owner along with the house tax in installments spread over some years with interest. This amount should be recovered as a loan repayment and not in the form of a tax. To meet this lacuna an amendment in the Municipal Act is needed. However, there are three alternative situations which require to be negotiated before the operation can be effected:

(a) where the owner is the occupier with exclusive use of the latrine;

(b) where the tenant has exclusive use of the latrine; and

(c) where the latrine is shared by two or more tenants along with or without the owner.

The owner will hardly be prepared to make any investment in situations 'b' and 'c', particularly when the tenancy is covered by the rent control legislation. However, the Rent Control Act in many states provides that if any development is carried out by the owner, he may increase the rent, charging interest on the works carried out by him. If the municipalities make conversions compulsory, the owners are entitled to enhance the rent. The way out for situation 'C' is that as per National Building Organisation's guidelines, one latrine seat is meant for 10 people in a household; hence if the number of people is higher than 10, two seats should be provided and a latrine for each family should also be proposed. Many municipal bye-laws provide for notice to be given both to the owner and the occupier, and in the event of non-compliance by the owner to carry out any works, the municipal authority itself can execute the work and recover costs from the owner and/or occupier according to provisions in their bye-laws. The municipality can, however, also permit the occupier himself to execute the work and deduct the cost from the rent due.

Legal Provisions and Their Enforcement

The "Employment of Manual Scavengers and Construction of Dry Latrines (Prohibition) Act, 1993 has been enacted, as passed by the Houses of Parliament – Lok Sabha on the 13th May 1993, Rajya Sabha on the 14th May 1993 and has been assented to on 5th June 1993 as Act no. 46 of 1993.

It would, however, be necessary to frame special bye-laws for compelling owners and occupiers to adopt waterseal latrines and prohibit the use of conservancy types. The Municipal Acts do provide for compulsory connections to sewers where they are laid and available within a specified minimum distance of the property. But there are no specific provisions for: (a) prohibiting the use of dry latrines; and (b) construction of water-seal privies in unsewered areas.

Surveys conducted in many towns indicated that even with the prevailing and binding laws mentioned above, in most cases of sewered towns even 50% of the houses have not been connected to sewers.

Mere legal provisions are not enough. The municipal administration has many laws, rules and bye-laws to enforce changes it deems fit in the interest of latrine programme. The local authorities have ample powers in all municipal Acts to compel the house owners to take sewer connections where underground sewers are provided. But it has not been possible to get a large number of water-borne latrines connected to sewers in areas where such systems exist. Prosecution involves long delays and nominal fines are no deterrent. Therefore, it is for the municipalities to modify these bye-laws to suit their requirements to enable their Executives, Health Officers and Municipal Engineers to enforce them. This calls for both administrative and political will.

The failure in enforcement is partly due to the absence of a positive approach to solve the problem. One can be prosecuted for committing nuisance, but it is equally important to tell the individual where to go for necessary facilities and to support a programme of public amenities. Enforcement programme must be backed by a systematic campaign of civic education by local bodies. While coercive powers are necessary to ensure eradication of conservancy privies, they require support by vigorous positive action to provide technical, financial and organisational infrastructure. Education and motivation are other important ingredients for the success of the programme.

Training Aspect

The low-cost pour flush latrine for unsewered areas has established its credibility and has become an operational part of the sanitary construction. And yet this low-cost technology requires all the ingenuity and expertise to provide for precision in construction and competence in supervision to guard against defective construction and faulty practices in the use and maintenance of this relatively cheap sanitary device.

It calls, therefore, for mounting a well planned training programme to create awareness and motivation amongst the engineer administrators, supervisors, construction workers, municipal staff and extension workers to appreciate that the solution to sanitation and liberation of scavengers in the country lies for the moment in the low cost sanitation and not high cost sewerage system.

Agency for Implementation

The two authorities which are directly involved in the implementation of the conversion programme are:

i) The engineering departments of state governments incharge of water supply and sanitation; and

ii) Engineering departments of the local authorities who have taken the responsibility of the implementation of the programme.

Both these authorities either get the work done by taking the assistance of reliable NGOs or the contractors. It is not possible to get responsible and well-established contractors to carry out this type of work. Their main indifference is because of the following : (i) this type of work provides a narrow margin of profit, (ii) it is very time consuming, being dependent on the convenience of the householders, (iii) it is scattered throughout the town and the design of pits changes with the soil conditions and the number of users, (iv) the latrines are located mostly in narrow lanes, which are difficult to approach and pose construction problems. In fact each household latrine is a project by itself. All these entail delay and lack of continuity Experienced NGOs are few and most of them have neither the requisite knowledge and manpower of creating awareness and carrying out motivation, nor have any sociological and technical back up. They have no experience in preparing a project proposal for submission to the local authorities. The local authorities have to employ their staff to get a project proposal prepared for submission to the Government. Since the staff of the local authorities is busy in various other obligatory duties, the preparation of the project gets delayed and implementation cannot be started as scheduled. Difficulties are experienced in house to house survey of the dry latrines and in creating motivation for the house owner to apply for the latrine conversion. The house owner also needs an agency which will take over the entire responsibility on behalf of the owner for compliance of all the formalities to secure necessary grant and loan. The agency has to submit to the local authority the agreement in standard form signed by the house owner. It has to keep a close liaison with the local authority to get the application processed and all formalities completed after individual verification of the work. The house owners also prefer to have a guarantee from the agency for the proper functioning of the latrine for some period and a contact office or person in the town where he can seek help of the agency in rectifying any problem that may be faced in future. It is impossible to find a suitable contractor and proper NGOs which would care to take all these responsibilities. Thus, finding a suitable agency also poses a constraint for the implementation of the programme.

Health Concept

In many areas the health concept is not understood by the people. Its implications in respect of liberation of scavengers are least understood. In these areas the people's psychology may be to get rid of the scavenging system and the nuisance value of the bucket/dry latrines. Therefore, area-specific programmes have to be developed on suitable lines so that the end objective is achieved.

However, there are areas where people have knowledge about primary health and sanitation. But even they would like to opt for the improvement of drinking water supply programme rather than the conversion/construction of the dry/bucket latrine programme. The approach here will be completely different and the authorities should first be briefed about the relative importance of safe water supply and sanitary disposal of human excreta. The importance of sanitation and the hazards caused by the bucket latrines is not realised by many of the local authorities and this adversely affects the implementation of the programme.

It is, therefore, necessary to decide upon the strategy which would facilitate successful implementation of the programme and achieve the set target. For this purpose the target that may have been decided upon has to be arrived at, which from all considerations is attainable. The success will naturally depend on the infrastructure i.e. administration, trained personnel, preferably with field experience, availability of funds and their timely release and peoples' participation through motivation, education, awareness campaigns etc. The target has to be demand-based and not number-oriented. The implementation of the programme suffers a setback where there is no response from the beneficiaries due to absence of felt need.

People's Participation

The need of people's participation for a successful programme of this type cannot be overemphasized. Experience has shown that in the absence of such participation the beneficiaries consider that the programme has been thrust upon them. Continued contacts with the community, in particular with the leaders of the area, through regular visits and house-to-house approach by women motivators to explain the need of the system and its benefits to the women, educational programmes and construction activities in the area under consideration will create the requisite local interest and will enable people themselves to take a lead in the sanitation programmes. In any case, follow-up visits to see that the latrines are properly used and looked after are very essential and should be ensured while planning such an activity. Absence of follow up and avoidance of it has resulted in many cases of choking of the system resulting in non-use of the latrines. People get apprehensive about the functioning of the system which has resulted in a setback in the further implementation. Follow-up helps in answering usual questions of the people like smell from latrines, breeding of flies/mosquitoes from latrines, safety of the latrines and the pits.

Traditional thinking which still regards pollution of the home by in-house or adjacent toilets can be modified only through proper education and motivation leading to effective community participation.

Selection of Priority Areas

Deciding the areas of priority and the selection of groups for providing the facility of this programme, the usual approach has been to consider the weaker section without considering their response and need. There are many instances where the pour flush latrines are not being used and the newly built latrine cubicle is used for other purposes like storage etc. This shows that the selection criterion should not be income-based. The facility should be made available to those who consider it a 'felt need' and are prepared and can fulfill all the conditions laid by the authorities. Among those who are classified under these categories, the poorest of the poor or the EWS of the community should be given the highest priority. This will facilitate implementation which often gets bogged down in getting the willingness of the beneficiaries to have latrines and also pay for the loan charges; in finding a suitable site for the facilities within the premises; in providing cooperation and, above all, in using and maintaining them. Another view is that only such areas should be selected where the sanitation is poor due to presence of bucket latrines or where the open drains are used by the inhabitants for defecation in absence of household latrines or properly maintained community latrines. In such cases by converting/constructing latrines, it will be possible to make a definite improvement in the environment and the health of the people, besides other social benefits. This will have an important effect on the people and there will not only be willing acceptance of the programme but better participation of the beneficiaries. However, nothing should be thrust upon the people. They should be motivated and educated and encouraged to show their keenness to accept the facilities. Implementation in such areas will thus be relatively easier for spontaneous people's participation.

Technology

The technology of two-pit low-cost pour flush waterseal latrines, as ideally suited to Indian conditions, is already accepted. Many of the implementing agencies, however, do not have clear conception of the technology and the need for its adoption. This lack of knowledge at times holds them back in explaining to the household the various aspects of the component of the system. Many households insist on using the conventional pans and traps. The reasons why specially designed pan and trap are required need to be explained to them. For want of proper education and awareness, many clients hesitate to use glass fibre pans, which have been approved by the Indian Standard Bureau and are being widely used for their durability. There also are situations where technological constraints, namely, high sub-soil water, pollution problem and lack of space dissuade the clients. It is, therefore, suggested that the authorities should have alternative technology to fit in such circumstances. At times the implementation agencies fail to explain the adoption of various designs required due to site conditions. They are at times not prepared to accept the design suggested by the authorities, as they have seen a different design which might be suitable in a different situation. Some implementation agencies insist on the location of the latrine and the pits at a particular site, which they feel will be more suitable. The constraint in space at times requires a different design or even shifting the siting of the pits to which the clients are not easily agreeable. This results in considerable delays and even higher cost.

Many clients are unaware about possible pollution from leach pit latrines. Apart from the possibility of pollution of ground water sources and water distribution pipes of the householder from his own leach pit latrine there is the possibility of polluting the neighbour's ground water sources and water distribution pipes. The elaborate investigations carried out on this aspect by the All India Institute of Hygiene and Public Health, Calcutta have conclusively demonstrated that the possibility of pollution flow arising out of pit privies is minimal and the system can be safely adopted in most soil conditions provided certain precautions are taken in relation to the ground water sources and water distribution pipes. Many of the implementing agencies and the beneficiaries are unaware of such possibilities. Thus the authorities have to take steps and make special efforts for explaining these.

Lack of delivery system for design, materials and resources can thus create difficulties in implementing the programme at a nation-wide level. The deficiencies have, therefore, to be made up to strengthen the delivery system.

Financial

The funds made available should be transferred to local authorities directly by Government of India, although the projects are submitted by the municipalities to the Government of India through the State Government. It has been experienced that State Governments keep the funds released by the Government of India and do not transfer it to local authorities promptly. This results in delay in the implementation of the programme and puts a strain on the local authority.

Another factor which needs attention is the release of fund for the project. Depending upon the estimated cost of the project and availability of funds, the period of completion of work should be assessed at the time of the sanction of the project. Funds needed for the first year programme should be released in one lot, because funds released in May-June are utilised after September due to the intervening rainy season when construction of latrines with leach pits at desired pace is not possible. The working period is usually from October to June and the carry over funds released in the preceding financial year are normally utilised from April to June. If the funds needed for one year are released in installments, the work cannot be planned. It has been experienced that at times, construction has to be stopped due to want of funds. It is desirable that as soon as 75% of the funds released in the first year are utilised, the funds needed for the second year should be released in one lot. Similarly funds for the third year should be released as soon as funds in full are released in the first year and 75% of second year are utilised. For utilisation of funds, the certificate of the local authority should be accepted.

Another factor which impedes the progress of the programme is that there is a time gap between the preparation of the project having cost-estimates and the actual availability of funds. The projects are sanctioned for the cost as per cost-estimates and by the time the funds are made available there is a gap of 2 to 3 years and the prices escalate in the meantime. This results either in the curtailment of work to the extent of available funds or part payments are made, resulting in delayed completion of the entire project. The local authorities also take time in making out a case for the sanction of additional funds and also in making payment to the construction agency for the escalated cost. Sometimes if there is no stipulation in the agreement to regulate costs, long drawn arguments continue, resulting in delays in completion of the assigned works.

Chapter - X

CONCLUSIONS AND SUGGESTIONS

On the basis of this comprehensive study on environmental sanitation and eradication of scavenging, several fresh insights have been gained and many significant conclusions have emerged. In the light of these, the author has offered a large number of suggestions to be given thoughtful consideration by appropriate governmental, non-governmental and international bodies interested and engaged in the gigantic task of improving India's environmental sanitation scenario by eradicating the need of manual scavenging and by liberating the scavengers condemned to carrying out this hateful occupation.

This chapter presents the summary of these conclusions and suggestions in two parts. Part-I presents some key-conclusions, and, Part-II presents the author's own suggestions, many of which he has already formally communicated to various concerned groups in a number of high-level official expert committee meetings in which he has been invited to participate on numerous occasions.

PART – I

The Problem of Environmental Sanitation

The problem of environmental sanitation has assumed considerable importance throughout the world, in all nations, both developed and developing. Crying concerns have been voiced in national and international fora regarding the state of environmental sanitation. In fact the realisation of the danger and damages caused by environmental pollution led to the acceptance of the necessity of environmental sanitation.

With the advancement of industrial/technological culture, the problem of environmental pollution has assumed alarming proportions and the situation is getting more and more critical. The growth of heavy industries and the concentration of human population in cities and towns has been responsible for increasing the degree of environmental pollution. In the urban habitats the human waste and heaps of garbage are important factor in increasing environmental pollution.

The growing realisation of environmental pollution and apprehensions of its dangerous consequences have been responsible for the genesis and growth of the idea of environmental sanitation. Environmental pollution is being regarded as the problem, and efforts for environmental sanitation are considered to be the possible solution.

With an increased stress on environmental pollution the problem of health and sanitation has also attained high importance in modern times. A number of international agencies such as the WHO, UNICEF and UNDP are working for improving health and sanitary conditions in most of the developing countries of Asia, Africa and Latin America. These and other national agencies, in respective countries, have realised the importance of good hygienic conditions as part of the qualitative development of human population. The programmes for the improvement of water supply and provision of an adequate system of human waste disposal have been taken up in different countries and discussions on various procedures and techniques are taking place at a number of seminars and conferences being organised at regular intervals.

The idea of environmental sanitation is closely linked with the idea of the eradication of scavenging. The safe disposal of nightsoil is regarded as the vital means to achieve the desired end of environmental sanitation. The need of environmental sanitation has developed to a great extent, out of the threat of environmental pollution caused by the system of manual scavenging and unhealthy and unsafe disposal of nightsoil.

Eradication of scavenging is being viewed as 'functional' not only from the point of view of improving ecological conditions and environmental sanitation, but also from the point of view of improving the degree of respect for human beings. Scavenging is a social curse and an abhorrent evil. It is against the notion of human dignity and against the concept of social justice. The work itself is degrading and dehumanizing. The continuation of service latrines and employment of scavengers to handle and carry buckets of nightsoil is a ghastly reminder of the hypocrisy of our social, cultural and political life. Thus, the campaign for the eradication of scavenging has a dual role to play in the Indian society : firstly, the restoration of human dignity to scavengers, and secondly, the improvement of the ecological conditions and environmental sanitation.

The Grim Scenario of Environmental Sanitation

Environmental pollution in its multiple and menacing forms has assumed the dimensions of a major crisis in India. Hence, prevention of environmental degradation has become a matter of serious concern in the nation's agenda of developmental priorities.

The issue of environmental sanitation is closely linked with the problem of environmental pollution, and, indeed, is a part and parcel of what ecologists and environmentalists call the "endangered human habitat". In fact, the realization of the damages caused to environment by an alarming prevalence of insanitation in cities and towns, has led to the acceptance of the need and significance of environmental sanitation. The awesome environmental insanitation scenario, especially in cities, has helped planning and promoting a holistic plan of environmental sanitation.

The impetus for the improvement of the country's sanitation scenario has been provided by the United Nation's call for designating a whole decade as the International Drinking Water Supply and Sanitation Decade (1981–1991). Accordingly, the Government of India started the National Water Supply and Sanitation Decade programme from April 1, 1981.

India's case of environmental sanitation is much different from the developed western countries. In the western world the primary focus of attention was reducing environmental pollution and thereby improving environmental sanitation. In India, the basic issue that engaged the attention of social reformers, policy planners and programme implementation functionaries, has been the eradication of scavenging –an obnoxious social evil. Environmental Sanitation is a later addition emerging out of the realization that the aim of environmental sanitation could hardly be achieved until scavenging is eradicated and scavengers are liberated. Hence, the eradication of scavenging emerged in the Indian society as a strategy for liberating scavengers, as also a measure of improving environmental sanitation. It is for this reason alone that a co-axis was formed between liberation of scavengers and environmental sanitation. This coaxis exists in the form of the programme of eradication of scavenging.

After independence a number of steps have been taken in different parts of the country to promote the cause of environmental sanitation and to speed up the implementation of the schemes for the eradication of scavenging and the liberation of scavengers. Several efforts have been made to conduct a variety of experiments for initiating viable projects for the improvement of environmental sanitation through low-cost sanitation programmes.

The schemes and projects launched with a lot of fanfare and fairly substantial financial allocations envisaged the emergence of a 'sanitation movement' in the country with increased accent on public health measures, entailing a package of programmes for safe disposal of human waste material. As the professed purpose was to launch a comprehensive sanitation drive in the country, the concerned international bodies (WHO, UNICEF, UNDP) also chipped in and offered handsome monetary and technical support.

Sanitation Status

The rhetoric notwithstanding, the nation's sanitation scenario continues to be too grim, despite about fifteen years of the observance of a much publicized sanitation decade. The facts, as mentioned earlier are both frightening and demoralising. These facts indicate that despite many past and present efforts the achievement of the country in improving the sanitation has not been significant.

Out of our total population of 1027 million (2001 census), 742 million (72.2%) in rural areas and 285 million (27.8%) in urban areas, 303 million people (29.5%) have access to adequate sanitation: 129 million (17.4%) in rural areas and 174 million (61.0%) in urban areas. This leaves about 724 million (70.5%) people who have no access to adequate sanitation: 613 million (82.60%) in rural areas and 111 million (39.0%) in urban areas.

There are about 13 million (6.9%) service latrines in the country, out of which 6 million (4.5%) are in rural areas and 7 million (13.0%) in urban areas. These latrines are dry type of latrines where human excreta is removed by scavengers.

63.06% of the total number of households in the country do not have latrine of any kind within the house: 78.01% in rural areas and 26.03% in urban areas.

In respect of the sanitation status in urban areas, the census of 2001 has revealed that the toilet facilities in the urban areas are available to 74% of the households which also include about 70 lakh bucket privies. The number of households having water closets in the urban areas is 25 million (46.1%) since sewer lines have been provided only in 232 towns and that, too, partially. 26% of the people in urban areas of the people in urban areas resort to open air defecation. It is the poor, particularly women and children who suffer most owing to the absence of toilet facilities in their houses.

The magnitude of the problem of sanitation, i.e., the lack of safe and hygienic toilets in rural areas is stupendous, as per the 2001 census. There are only 22% of the households which have toilets and the remaining population of 78% households goes for open defecation. Considering the fact that 78% of the country's population lives in rural areas, the position, to say the least, is extremely unsatisfactory.

Low sanitation coverage in rural areas is primarily due to widespread poverty, illiteracy and lack of knowledge about the elementary principles of health and public hygiene. The people are not aware of adverse effects of open air defecation which is responsible for a variety of infections and diseases. People even now prefer to go for open defecation rather than toilets in their houses. With the passage of time, there has been slight change in the outlook but by and large the mindset still persists. The facility of toilet is not a felt need and people prefer to spend on food and clothes. The problem is further compounded owing to the small size of the mud built or semi-brick structure houses where there is no enough space for constructing a household toilet.

The absence of toilet facilities in schools in rural areas has adversely affected girls' education. The parents do not send their girls as there is almost complete absence of toilet facilities. While the boys can go for open defecation, the girls cannot. This has resulted in low enrolment of girls and the number of dropouts has also increased.

Safe disposal of human excreta is the most challenging problem, both in rural and urban areas. The lack of modern sanitary system (toilet facility with in-built system of disposal of human excreta), health-hazard, free drainage system, unavailability of garbage disposal facilities and the scarcity of safe drinking water continue to be the most pressing problems. Amongst these, the problem of open-air defecation in both villages and cities is truly a matter of national shame. In the urban communities of the country, the problem of environmental pollution is very grave due to the ever-increasing excessive population pressure, development of large slum areas, inadequate facilities for the disposal of heaps of garbage and cleaning of drainages, unplanned building constructions and huge amounts of industrial wastes. The gravity of the problem has further increased due to large and extensive dry latrine system with 1,31,20,867 service/dry latrines in the country. Attached to this is the problem of cleaning these latrines through manual scavenging done by a certain section of society. Thus, a painful paradox persists. While on the one hand a gigantic leap in the field of science and technology and has advanced both economically and technologically, there, however, exists the centuries old practice of manual handling of human excreta.

While from centuries onward, the country's culture places a premium on personal cleanliness, there is a widespread disregard for keeping the habitat clean and health-hazard free. The provision of a latrine has been traditionally ignored as an integral part of a house in the Indian home. Returning human excreta direct on to the soil by using the open field for defecation was a matter of habit and convenience before the advent of modern civilization. Rapid urbanization brought home the need for a latrine in the house. Even so the household privy has not found favour in many houses of the poorer sections of society. It is also common to find houses in towns and cities without any latrine, although these houses have provision of a living room and a bedroom. This is obviously because there is no 'felt need' or importance attached to have a latrine within the house. True that a household privy might not be feasible technically for lack of space, or for economic reasons, but in many cases there is a lack of consciousness to have one and to give it a priority and to consider that a house is really incomplete without a latrine.

In India, unlike other developed or some other ocial Service Organisation , developing countries, sanitation is not a priority area in developmental policy planning, and hence, the financial allocations for the purpose are meagre. The result is that there is nothing like a viable and vibrating programme of environmental sanitation, either in the Government's scheme of things or in the peoples' perception. More disconcerting is the fact that there is neither much governmental intervention, nor enough voluntary action to concretize the major concerns of environmental sanitation.

The Nature of Sanitation Movement in India

In the context of scavenging, the sanitation movement in India has been typical in character. Viewed as a collective mobilization for sanitation, it refers to physical health and hygiene for removing or at least reducing environmental pollution through proper disposal of nightsoil. But along with it, the sanitation movement in this context refers to the liberation of scavengers. Thus, in respect of scavenging, the idea of environmental sanitation has been tagged with the liberation of scavengers.

Initially, the movement had its genesis in the sporadic and isolated efforts for improving the condition of scavengers and for liberating a particular hereditary group or caste or sub-caste from obligatory work of a demeaning and sub-human nature. Thus, originally the sanitation movement in India started in the form of a movement for liberating scavengers or at least for improving the social condition of this occupational hereditary group. It is obvious that the idea of emancipation of scavengers refers to a change in normative patterns, value orientation and, thus, a change in social order or social structure. Like all other movements, this movement was also preceded by a number of isolated events, efforts and struggles. All these have redefined the perspectives on environmental sanitation and eradication of scavenging. The redefined perspective contributed to the development of an ideology, which was shaped and crystallized under the charismatic and able leadership of Mahatma Gandhi. In this way, the foundation for the emergence of this sanitation movement was laid and the isolated struggles and efforts made in this direction led to the intensification of a clean environment campaign in all parts of the country. The movement gradually produced a sense of solidarity and the leadership provided orientation towards action. Thus the movement for the liberation of scavengers started in the country and is continuing even today at different levels and in different forms. The dimension of environmental sanitation was also added to this movement, since environmental sanitation appeared to be linked with the movement for liberating scavengers.

The movement now entails efforts for improving sanitary condition and the social status of scavengers who had been held in contempt for centuries. Eradication of scavenging has, therefore, been recognized as an extremely significant dimension of environmental sanitation. So long as this evil persists, the environment will continue to be polluted. Hence, eradication of scavenging and environmental sanitation are inextricably interlinked. In the context of Indian society, liberation of scavengers strikes at the very root of the centuries old institution of untouchability embedded in the caste-ridden social structure.

The sanitation movement that was initiated by Gandhiji in Calcutta in 1901, and formally launched by him at Sabarmati Ashram in 1918 and later on taken up by his followers (Vittal Phalke, Balcoba Bhave, Jagat Ram, Appa Saheb Patwardhan) in Gujarat and Maharashtra created a climate conducive to the development of programmes of rural sanitation in the country. In the following years a number of experimental works and programmes of action in respect of rural and urban sanitation were taken up with the assistance of Governmental machinery and some international agencies. These experiments served as landmarks in the history of the sanitation movement in the country.

The sanitation movement in India entered a new era with the appearance of Sulabh Shauchalaya Sansthan in 1970 (now Sulabh International Social Service Organisation). The entrance of Sulabh International in the arena of sanitation movement gave a new turn to the programme of environmental sanitation, which until 1970 was mainly concerned with rural sanitation. Sulabh International initially took up the programme of urban sanitation and the liberation of scavengers working in cities and towns. The organisation chose urban sanitation as its principal area of operation in view of the fact that the problem of environmental pollution is very grave and alarming due to the ever-increasing and excessive population pressure, development of sprawling slum areas; inadequate facilities for the disposal of garbage and cleaning of drains, unplanned building constructions, huge stockpiles of industrial waste, and vast number of dry/bucket privies requiring the services of about half-a-million scavengers. Sulabh International devised a low-cost sanitation technology which obtained soon an enormous amount of national and international recognition. Along with achieving considerable success on the front of converting household dry service latrines into low-cost sanitary latrines, the Organisation has also started the programme of public latrines through building Sulabh Community Toilet Complexes at different public places in cities and towns all over the country.

The appearance of Sulabh International Social Service Organisation – a pioneering voluntary organisation, basically dedicated to the eradication of scavenging and the liberation of scavengers, has proved to be a significant milestone on the road to environmental sanitation. The Organisation has emerged as the most popular and dynamic agency in the field of environmental sanitation in a relatively short period of 40 years. The achievements of the Organisation are spectacular and have been highly appreciated by all the concerned national and international organizations. The initiatives taken by Sulabh International and the technology of low-cost sanitation advanced by it, have been hailed as an outstanding innovation in combating the menace of environmental pollution in underdeveloped and developing countries.

Appropriate Technology

Sewerage System

No doubt sewerage is the ideal system for disposal of human waste as it takes care of both human excreta as well as domestic waste water from bathrooms and kitchens but it has its limitations. Although sewerage was introduced in India about 130 years back and the entire Five Year Plans allocations for urban sanitation have almost been spent on sewerage system, either on-going or new, yet there are nearly 232 towns and cities only, out of 5161 towns with sewerage. None of them, however, covers the entire municipal city areas, leave alone the adjoining suburbs included in the municipal limits. The percentage of households having water closets in the urban areas is only 46 as per 2001 Census. Besides, in most of the towns due to paucity of resources, untreated sewage is being discharged into rivers polluting them heavily. Even on the streets where sewers have been laid, houses have not been connected inspite of municipal laws making such connections compulsory, thus insanitation continues.

Sewerage involves enormous investment in laying sewer lines, treatment plants and related works, besides high operation and maintenance cost. In developing countries like India neither the government nor the local authorities or the beneficiaries can bear the capital or the operation and maintenance cost. Moreover the sewerage requires skilled persons and good management to operate and maintain it. It needs considerable quantity of water to function properly. For flushing toilets connected to sewers almost 6 times more water is required than that for pour-flush system. Since in most of the towns there is scarcity of water, and if sewerage system is desired to be introduced, water supply will need augmentation involved more capital investment. Moreover since all the houses on a street are not connected, the household waste water along with waste from the streets is led into the sewers through gully pits. Discharge of such waste and soil and insufficient water due to lack of house connections is resulting in chokage of sewers and making many of them non-functional. In a number of towns even big sewers get choked due to lack of inflow of sewage as all the houses are not connected to sewers. One of the reasons for non-connection of houses to sewerage is that laterals and branches are not laid in every lane; hence cost of connection becomes very high for many houses due to laying of large length of sewer to connect the house.

Sewage systems are built to protect public health but badly managed sewers can become a serious health hazard. There can be serious outbreaks of water borne diseases from:

i. River pollution because of sewage outfalls;

ii. Groundwater contamination because of leaky sewer lines;

iii. Contamination of piped water supply systems because of leaky sewer lines leading to infiltration of pathogens into drinking water pipelines, especially when they do not have water, which is the case in many cities in developing countries as they cannot provide water round the clock; and

iv. Sewage backflows because of badly maintained and

blocked sewers or because of increasing use of non-biodegradable materials like plastic bags.

The use of sewer systems would have totally destroyed the aquatic ecosystems in the developing world, posing enormous threats both to public health and aquatic bio-diversity. In India, we do not even have to look a few years ahead. We already see the signs of this hydrocide. Literally, no small or medium river today is clean. Every river that passes through a city or a town becomes stinking sewer.

The sewerage system needs highly qualified technical personnel for designing, construction, operation and maintenance. Developing countries like India cannot afford it due to exorbitant capital, operation and maintenance costs. The local authorities on account of lack of resources are unable to bear the entire O&M costs.

All this makes water sewerage a waste disposal paradigm that is extremely expensive because of its high economic, environmental and public health costs.

Septic Tank

Although septic tank was developed about 400 years back and introduced in India about 100 years back, yet even in urban areas about only 20% of the houses have septic tank toilets. The septic tank has some advantages like flexibility and adaptability to a wide variety of individual households. It is suitable for scattered houses with space for septic tank construction. However, the limitations of septic tank outweigh some of the advantages which have been mentioned above. Some of the limitations are mentioned below:

i. Septic tank requires more space and thus in congested areas and in small houses where space is a constraint, construction of septic tank latrine is not feasible.

ii. It is costly and it is beyond the reach of most of the people.

iii. It has many drawbacks and operational problems. It has to be cleaned after 1 or 2 years. The wet sludge taken out is a health hazard and at times gets mixed up with fresh night-soil and has obnoxious smell. The sludge, therefore, has to be discharged at a safer place. Emptying of septic tank is not easy and is a burden on the households and the municipalities.

iv. The quantity of 14 litres of water is needed for flushing.

v. For cleaning of septic tank and disposal of sludge, services of scavengers are needed. The Govt. of India has enacted a legislation to ban scavenging in such a situation.

Sulabh Shauchalaya (Twin-pit Pour-flush Water-seal Toilet)

Sulabh Shauchalaya is the most appropriate technological option to serve as an alternate to bucket privies and to stop open air defecation. It can be constructed in the most congested areas of cities and towns. Sustainability and replicability are two important issues which should be considered while choosing an appropriate technology. Sulabh Shauchalaya fulfils these criteria fully. It is socio-culturally acceptable, affordable, easily available and users are able to operate and maintain it easily. It is an indigenous technology and the toilet can easily be constructed by local labour and material. It is scavenging free and requires only 2 litres of water for flushing. When one pit is full, excreta is diverted to the second pit. In about two years' rest period, the sludge gets digested and is almost dry and becomes safe for handling. Digested sludge is a good manure and soil conditioner. Sulabh Shauchalaya can be designed with different specifications and use of different materials in varied costs without compromising the design principles. Thus the need of families with different incomes can be met easily. The Sulabh technology can be implemented easily in different eco-cultural and economic set ups and in different physical, geological and hydrogeological situations. The latrines seat can be squatting type or pedestal type depending upon the defecation habit of the users. When Sulabh Shauchalaya technology is compared as obtained this is most suitable cost-effective technology which will ensure wider coverage with the least investment. It fulfils all the criteria of an appropriate technology.

The Scourge of Scavenging

Several historical accounts of ancient vintage bear testimony to the fact that scavenging in India, like some other countries, began with the rise of urban civilization and persisted in all ages – vedic, ancient, medieval and modern. A particular section or class in society was assigned to do this work on hereditary caste-basis. Throughout the ages scavenging has been considered to be an unclean profession, degraded and despised. The so-called 'inferiors' and weak people in the caste-hierarchies were assigned to do this dirty work. Scavenging was made the duty of a specific group, which was regarded as 'untouchable' and which occupied the lowest position in society. Being 'Mlechhas' or 'Chandals' they were regarded as impure and polluted and were condemned to do dirty jobs like sweeping and scavenging. In ancient times, these 'unclean people' had been subjected to myriad tyrannies. Being social outcastes, and being downtrodden – socially, culturally, economically and politically – they have been subjected to indignities, atrocities, violence and victimization.

The Plight of Scavengers

While no reliable data are available regarding the exact number of scavengers engaged in the occupation of scavenging, some official estimates (questionable though), churn out figures which outline the magnitude of the problem; nearly half-a-million scavengers are engaged in demeaning and dehumanizing practice of manual handling of human excreta.

All available writings on scavenging and scavengers paint a pathetic picture of the social status of scavengers and the existential reality of their life and living. Being members of a socially ostracized class/caste, scavengers are still being treated as 'impure', and consequently, ritually avoided. Like 'pariah' groups they are treated as inferior and often regarded as less than human. They live on the fringe of society and are excluded from social contact, though their services are accepted as highly valuable. They are backward and often oppressed, not because they are racially inferior to others, but simply because they have been assigned the lowest of the low status in the social hierarchy. They have been treated like a dreaded contagious disease calling for a disdainful distance and minimal social contact. Besides social and educational backwardness, scavengers are economically backward too. Because of their social, economic and educational backwardness, most scavenger families have been deprived of taking full advantage of the privileges meant for the welfare of the dispossessed. Politically too, scavengers are little organised as compared to other Harijan sub-castes, the dalits and the disadvantaged. While in recent years there has been some change in the traditional occupational structure of the scavengers, a substantially large majority of them is still given to traditional hereditary occupations. In summing up the plight of scavengers, Mahatma Gandhi – the Father of the Nation – made remark in 1927 that still holds true: "Socially they are lepers and economically they are worse than slaves".

National Concern on Scavenging

National concern on scavenging was expressed in large numbers of Reports of the Commissions and Committees right from the days of Independence. The concern was more forcefully voiced in a Report of the National Labour Commission on the working and service conditions of sweepers and scavengers (1968-69). With extreme anguish the report observed: "That a section of humanity should be condemned to such degrading condition is a slur on our culture and a disgrace to our society".

Many great national leaders expressed their concern (as also indignation) on the situation of scavengers. Mahatma Gandhi said: "I may not be born again, but if it happens, I will like to be born in a family of scavengers so that I may relieve them of the inhuman, unhealthy and hateful practice of carrying nightsoil". Smt. Indira Gandhi said: "Everybody is conscious of this fact (inhuman system of carrying and handling human excreta manually), and we have discussed this many times. There is no doubt that this is a disgrace and it should be eradicated as early as possible". Rajiv Gandhi said: "The continued practice of scavenging is the most undesirable form of social derogation".

Reform Measures

Many social reformers have in the past voiced serious concern on the plight of scavengers, and even worked for their liberation from social and economic bondage. Their crusade for the liberation of scavengers did not produce much worthwhile results, simply because the very idea of the liberation of scavengers militated against the well-entrenched values of a caste-ridden society. Mahatma Gandhi too had a dream for the liberation of scavengers, but his efforts produced no tangible results, at least in his life-time.

After Independence, the Central and many State Governments started work for the liberation of scavengers, and appointed Commissions/Committees to suggest remedial measures. These Commissions/Committees did a splendid job and made many valuable recommendations for the consideration of the Government. International agencies like WHO, UNICEF and UNDP also chipped into serve the cause of scavengers' liberation from human bondage.

While there is no denying the fact that some significant work has been done in the post-independent India, the problem, without doubt, persists as a continuing disgrace to the nation. Many welfare schemes launched for the liberation of scavengers have not made any significant dent in the problem.

The evaluation studies made to assess the impact of the Governmental programme have revealed that the efforts fell short of what is really required to liberate the scavengers. The main weakness has been the absence of priority to the programme in the Government's planning process. The studies have also revealed that despite the emphasis by all political leaders and efforts made by Government since Independence, not much tangible improvement seems to have been made. The extent of scavenging in the country still continues to stand at an alarming point.

Official documents admit this fact. The Eighth Plan document avers: "Sanitation, both in rural and urban areas continues to be neglected... the performance continues to be unsatisfactory and the concept of 'total environmental sanitation', seems to have received scant attention." In regard to scavengers the document says: "It has been recognized that earlier efforts to discourage manual scavenging and liberate families engaged in this inhuman task have not produced satisfactory results." A widely published HUDCO document admits: "despite the fact that concerted efforts have been made to eliminate the dehumanizing practice of employing persons for carrying human excreta on their heads, or likewise, the practice still persists in many parts of the country. The problem of manual scavenging has been aggravated because the core concerns of the issue at hand have not been effectively dealt with i.e.; the conversion of dry latrines into sanitary pour-flush latrines has not been handled in any systematic manner."

It is a matter of concern that during years of existence of the scheme of Liberation of Scavengers (1980-81) with twin objectives of : (A) conversion of the existing household/ community latrines into pour-flush water seal latrines on a 'whole town approach' basis in selected and small and medium towns; and (b) retention of all municipal scavengers by assigning dignified occupations and rehabilitation of all private scavengers in alternative employment/occupations, not much tangible progress is visible. Only about 10.77 lakh dry latrines in 1534 towns have been converted and there still remain several lakh dry latrines to be converted. Although 490 cities/towns were taken up to make them scavenging free 650 towns out of 1534 towns have been made scavenging free.

The system of manual scavenging has contributed to pollution of environment in different ways. The accumulation of human excreta in the household latrines is a major source of pollution, this human waste remains exposed to the air and the gases coming out of night-oil and the flies and insects sitting on the night soil pollute the environment. The system of carrying headloads of night soil is also responsible for environmental pollution because the route through which the scavengers carry night soil becomes infected by the germs and flies sitting on the human waste. The system of bucket privies the environment remains stinking and the constant contact of flies with night soil gives birth to diseases like diarrhoea, dysentery, cholera, hookworm, ring worm etc.

The present system of manual scavenging is a major source of environmental pollution. The huge quantity of human waste thrown at the outskirts of the cities and towns is the central cause of environmental pollution, because it remains in direct contact with air. There is not only the problem of air pollution through huge deposit of night soil, rather other problems of health hazard also arise out of it. In view of these, efforts to reduce environmental pollution demand a check on the system of scavenging and cleaning of service latrines, carrying headloads of night soil and allowing huge deposits of human waste at a particular place.

The system of manual scavenging continues defiantly even after the introduction of septic latrines in urban areas. The cleaning of septic tanks is done manually and physically touch with night soil becomes inevitable. In such a situation, scavenging takes place naturally. At the time of opening and cleaning of the septic tank the surrounding is polluted because it is exposed to human waste. The night soil collected from the septic tank is indiscriminately disposed of, and it usually remains exposed to the air, causing pollution to the surrounding environment.

While the sewerage system available in the cities has reduced the demand of scavenging, it has, however, failed to check the problem of environmental pollution. The prevalence of sewer system has polluted the water of many rivers.

All the three major systems of the disposal of night soil, namely, dry latrines, septic latrine system and sewerage have failed to prevent pollution in the environmental and a large number of people are exposed to unhygienic and unhealthy environmental conditions. Thus, environmental degradation caused by human excreta is becoming increasingly serious.

The practice of manual handling of night soil still exists in large parts of the country despite continued efforts and best of intentions to eliminate it. Manual scavenging persists because of the prevalence of dry latrines.

The unabated use and the continued existence of dry/bucket privies are the principal culprits of environmental sanitation. The reasons are far too many. Lack of political will (giving high priority to the programme), indifferent attitude of the authorities, flawed decisions in regard to the selection of appropriate technology, behavioural attitude of scavengers and householders, legal provisions and their enforcement, paucity of trained manpower etc. have been responsible in some ways or other in slowing down the progress of several programmes of environmental sanitation, especially conversion of bucket/dry latrine programme and the liberation of scavengers from traditional, hereditary occupations.

Problems in Planning

Lack of Priority : Notwithstanding the emphasis by the political leaders, planners and administrators, the efforts made by the Government have fallen short of what is really required to liberate the scavengers. In fact, the absence of priority in the various plans has been the single biggest factor for which there are still 1,27,872 scavengers engaged in the demeaning tasks of manually handling human excreta even after 62 years of Independence.

The 2-Point programme, formulated first in 1972, to serve as a guide for fixing priorities of a planned programme of developmental nature, did not include 'sanitation' as an item; nor was it included in the 20-Point programme of 1982. The item was, however, included in the revised 20-Point Programme of 1986

Flaws in the Selection of Technology : The flawed planning in the selection of appropriate technology has hampered the attainment of desired targets despite heavy expenditure and almost full utilization of funds allocated for the purpose. For piquant reasons the planners provided more emphasis on laying sewerage system, at the expense of already proven and available low-cost two-pit pour-flush technology to be easily and effectively adopted in a variety of urban settings, where manual scavenging of human excreta is widely prevalent and where a large chunk of scavengers are engaged in a demeaning task of clearing and carrying night-soil as head load. The most unfortunate aspect of this flawed selection of technology has been that in the cities where sewers have been laid, city residents have mostly refrained from taking house connections to sewer for financial constraints. The result is that a large number of dry/bucket privies exist in all cities, necessitating the continuance of a much maligned system of scavenging. If the planners had set their priorities on the liberation of scavengers in the cities and towns in the country, entailing the conversion of bucket/dry privies to leach pit latrines and construction of new sanitary latrines where none existed.

Sidetracking of Critical Questions : The planners did not address themselves adequately to the most important single question: what is affordable, socially acceptable, easily available and technically feasible that the users would readily accept and be able to maintain and which the local authorities are institutionally capable of operating? Had the planners addressed themselves to the issue of on-site sanitation, i.e., twin-pit pour-flush waterseal latrines popularly known as Sulabh Shauchalayas, they would have found a satisfactory answer. A little more serious thought could have also provided the planners a more concrete measure of liberating the scavengers from the demeaning task of handling and/or carrying head-load of buckets containing human excreta.

Planning Process Being Devoid of Pragmatism: The planning process was entirely devoid of pragmatism in respect of laying down a fairly wide and viable sewerage system in class I, II and III cities. Had the planning been realistic and had the major allocation under urban sanitation been made for conversion/construction of latrines in the successive Five Year Plans, the country would have been able to tackle to a considerable extent problem of scavenging and scavengers, especially in metropolitan and relatively larger cities where the problem is more acute. Such a strategy of urban sanitation may have considerably improved the sanitation scenario of the cities. In doing the financial planning the planners lost sight of the fact that sewerage is a luxury and the system is not suitable for a poor country like ours. Sewerage is not only highly cost-intensive but also requires considerable expense in addition to skilled and trained staff for operation and maintenance.

Constraints in Implementation: The houses which have bucket or dry latrines or which do not have latrines, indeed, belong to that section of the people who can neither afford to get these latrines converted nor have necessary means to construct latrines in their houses. While the average urban householder with a bucket latrines looks upon it as an evil, he is, in most cases, unable to bear its conversion cost. Even if loans and subsidies are offered for conversion of bucket privies or, for the construction of two-pit pour-flush latrine (as designed by Sulabh International), the various procedural formalities are too cumbersome and time-consuming to scare away most such householders. Though the standard water-closet with connection to sewer is far more expensive, there is considerable skepticism about the design and working of the cheaper hand-flush toilet system. And then, there is yet another difficulty of arranging men and material for construction of the toilet. Another reason for the reluctance of converting the dry latrine on the part of the householder is the low-cost of maintaining and cleaning the dry latrines at some places. The municipal rates in most of the local authorities are very low for cleaning the family bucket privy, which discourage a changeover. If the local authorities give a substantial rebate on the scavenging/drainage tax to the owner who switches over to a water flush toilet, there will be a better response. In states where private arrangements are feasible and scavenging charges for the latrines are high, it has been observed that there is more eagerness and response from the households for the conversion of bucket privies.

The introduction of a new financing pattern based on income level of the people has created unnecessary confusion, as it is extremely difficult to identify to which category the house owner belongs. The new financing pattern, as the field experience indicates, has retarded the progress of the scheme for the conversion of old latrines and construction of fresh ones.

For most of the owners of the houses having bucket/dry latrines or no latrines, it is not possible to set apart their share in one lump sum, while they can meet it by paying in easy installments. The changed financing pattern, as already indicated, has retarded the progress drastically as the construction has been dependent on the availability of beneficiary's contribution, which is not forthcoming readily.

Another factor which impedes the progress of the programme is that there is a time gap between the preparation of the project having cost estimates and the actual availability of funds. The projects are sanctioned for the cost as per cost estimates and by the time funds are made available, there is a gap of 2 to 3 years and prices escalate in the meantime. This results either in the curtailment of work to the extent of available funds, or part payments are made, resulting in delayed completion of the entire project. The local authorities also take time in making out a case for the sanction of additional funds and also in making payment to the construction agency for the escalated cost. Sometimes, if there is no stipulation in the agreement to regulate costs, long drawn arguments continue, resulting in delays in completion of the assigned work.

Pattern of financial allocation and its transfer to appropriate local bodies is far from satisfactory. Funds made available should be transferred to local authorities directly by the Government of India, although the projects are sent by the municipalities to the Government of India through the State Governments. It has been experienced that the State Governments keep with them the funds released by the Government of India and do not transfer them to local authorities promptly. This results in delaying the implementation of the programme and putting a strain on the local authority.

Many of the implementing agencies, however, do not have clear idea about an appropriate technology and the need for its ready acceptance and adoption by the planners. This lack of knowledge, at times, holds them back in explaining to the householders the technical details and the advantages of low-cost technology.

The approach followed in deciding the areas of priority and the selection of groups for providing the facility under the programme has been faulty. The usual approach has been to consider the weaker sections as beneficiaries under the programme, without any regard whatsoever to their need or response. Since the selection is income-based, people with 'felt need' and encouraging response are beyond the overreach of the scheme.

The importance of sanitation and the incidence of health-hazards caused by the existence of bucket latrines has not been fully realized by many of the local authorities. The lack of proper realization on their part has adversely affected the implementation of the available programme of action.

The exiting legal provisions in the municipal laws until the enactment of the Employment of Manual Scavengers and Construction of Dry Latrines (Prohibition) Act, 1993, only provided for compulsory connection to sewers, where they are laid and available within a specified minimum distance from the house. After about sixteen years, the Act has not been able to discourage and/or deter people from the continued use of dry latrines. It is, therefore, obvious that mere legal provisions are not enough. Had that been so, a majority of the household water-borne latrines could have been connected to sewers. The real difficulty lies in launching successful prosecution in the cases of recalcitrant householders. The experience, however, involves long delays and nominal fines are no deterrent.

The failure of enforcement is also due to the absence of a positive approach in solving the problem. One could be prosecuted for committing nuisance, but it is equally important to tell the individual where to go for the necessary facilities, i.e. the provision of technical, financial and organizational infrastructure.

Attitude of scavengers themselves has been a strong deterrent in the conversion of bucket/dry privies, as also in their liberation. This has been primarily due to the following factors: (i) since scavenging has been traditionally accepted as a way of life, scavengers fear that their livelihood may be lost if bucket privies are abolished; (ii) majority of the scavengers suffer from a fatalistic outlook on account of their low level of education, with nobody around to educate and guide them about the governmental programmes for their liberation from manual scavenging. In this situation it is natural that the resistance to change has come from the sources within.

Contribution of Governmental, Non-governmental and International Agencies

Various governmental, non-governmental and international agencies have made a noteworthy contribution in furthering the cause of environmental sanitation and liberation of scavengers. Research Institutes at Singur, Najafgarh and Poonamalli have made their contribution by designing various types of latrines and undertaking several action programmes. In the same way various non-governmental (voluntary) organizations like Gandhi Smarak Nidhi, Harijan Sevak Sangh and Safai Vidyalaya have worked hard for improving the environmental sanitation as well as ameliorating the condition of scavengers. Their programmes and activities have been of pioneering nature and have definitely helped in creating a climate favourable to the promotion of the cause of environmental sanitation and the eradication of scavenging and liberation of scavengers.

Along with governmental agencies and non-governmental organizations, the international agencies like WHO, UNICEF, UNDP and World Bank have also played an important role in highlighting the critical issues of environmental sanitation and also in contributing in abundant measure to the country's campaign on the eradication of manual scavenging. These agencies have not only provided financial assistance, but have also given technical assistance for implementing various sanitation schemes launched by several governmental and non-governmental organizations in the country.

PART-II

General Suggestions

In order to tackle the problems relating to environmental sanitation and the eradication of scavenging and liberation of scavengers and their dependents from the hateful practice of manual scavenging, an urgent and speedy action in the following five areas is essential:

1. Assessment of the problem of manual scavenging in terms of the total number of scavengers, the number of towns and places where it is practiced.
2. Analysis of the full magnitude of the problem and its different dimensions, with a view to suggesting a time-bound comprehensive programme of action for the abolition of this abhorrent practice.
3. Introduction and initiation of several interventionist measures, including the spirited enforcement of the Employment of Manual Scavengers and Construction of Dry Latrines (Prohibition) Act, 1993.
4. Popularization of cost-effective technology for conversion of dry latrines.
5. Identification of agencies to tackle the problem in a well-coordinated manner.

In respect of these five areas, the following suggestions deserve serious consideration.

a) A broad assessment of the number of household having bucket/dry latrines can be made on the lines of house to house survey already carried out by UNDP/World Bank during 1980-84 in 211 small and medium towns of 21 States and union territories covering 15 lakh households.

b) The analysis of the full magnitude of the problem has to be attempted by taking into account the socio-economic, behavioural, cultural and environmental conditions of scavengers.

c) Conversion of all dry latrines into water-flush latrines for relieving the scavengers from in-human and undignified task of cleaning dry latrines and carrying human excreta manually.

d) House owners who have no latrines, but have space for construction of latrines, should be provided with sanitary latrines so that they do not construct dry latrines that require the services of scavengers.

e) Sufficient number of community toilet complexes with bathing and washing facilities may be provided for: (a) the floating population; (b) those who do not have home space for a latrine and (c) those who continue to resort to open defecation. These toilet complexes should be run and maintained on 'pay and use" basis, as is being done by Sulabh International, so that their maintenance is proper and self-supporting.

f) The facilities for the training of the wards of scavengers likely to be rehabilitated need enlargement, augmentation and upgradation so that they are suitably rehabilitated rewarding trades and occupations.

g) In order to combat the scavengers' resistance to change, there is an urgent need for a forward looking mass education policy on the part of the Government of India, the State Governments and the local bodies to make the scavenger community sufficiently aware of the series for financial assistance should be made aware of various avenues that are open to them for financial assistance and for training and rehabilitation in other dignified and rewarding occupations. Dedicated non-governmental social organizations should be called upon to undertake the task of guiding and educating the scavengers about concessions and facilities presently available to them.

h) An adequate number of specialized training institutions are required for scavengers and their wards to impart training in such trades and professions which have considerable demand and can be imparted to people of lower educational standards.

i) The institutions which are already engaged in offering training to the scavengers and their wards should also be entrusted with the tasks connected with their rehabilitation.

j) The rehabilitation programme should include establishment of shops, provision of necessary tools and machines along with raw materials and other consumables.

k) Need for a determined political will along with firm policy commitments, backed by effective decision making and implementation apparatus at the Central, State and local levels.

l) Establishment of a sound communication support system to propagate the idea of environmental sanitation and elimination of scavenging.

m) Involvement of voluntary organizations in the implementation and follow-up of the programme.

n) Every State Government must make adequate budgetary provisions for promoting the programmes of environmental sanitation and the eradication of manual scavenging, whilst according the programmes a high priority.

Specific Suggestions

Conversions of Dry Latrines into Waterseal Pour-Flush Latrines

1. For speed and ease of conversion of dry latrines, the present financing pattern adopted by the HUDCO has proved counter-productive, as it is difficult to identify the category to which the house-owner belongs. The experience suggests that this financing pattern has retarded the progress in the implementation of the scheme. It is, therefore, suggested that the financing pattern should be uniform, i.e. 45 per cent grant and the balance as loan for all categories. This uniform pattern was in vogue earlier.
2. In order to fulfill the commitment of the Government of India for the eradication of scavenging in the country, it is absolutely essential to eliminate beneficiaries' contribution. If the Government desires, the subsidy component may be further reduced, but the loan portion should be raised to meet the balance cost of conversion or construction.
3. The rate of interest on HUDCO loan should be reduced to 4% and retiring period increased to 20 years so that the installment for repaying the loan is within the affordable limit of the beneficiaries who mostly belong to EWS and LIG.
4. Construction of latrine superstructure should be financed on the same pattern as it is for latrines upto substructure level. It has been our experience that if superstructure is not built along with latrine, a large number of people do not construct it as they cannot

afford to do it. It results in non-use of the latrine. Thus the expenditure incurred on the construction of latrine upto sub-structure level becomes infructuous.

5. For releasing the funds, the HUDCO examines the financial position of the local authority. It is well known that the financial position of most of the local authorities in India is very weak. Hence the HUDCO should release the funds on the guarantee of the State Government.
6. The funds should be transferred to local authorities directly by the Government of India, though the projects may be submitted by the municipalities to the Government of India through the State Government. It has been experienced that the State Governments keep the funds released by the Government of India in 'Ways and Means' account and do not transfer it to local authorities for 2 to 3 years. This results in the stoppage of work.
7. Depending upon the estimated cost of the project and availability of funds, the period of completion of work should be assessed at the time of the sanction of the project. Funds needed for the first year programme should be released in one lot, because funds released in May-June are utilized with leach pits at the desired pace is not possible. The working period is usually from October to June and the carry-over funds released in the preceding financial year are normally utilized from April to June. If the funds needed for one year are released in installments, the work cannot be planned. It has been our experience that at times, construction has to be stopped due to want of funds. It is also suggested that as soon as 75% of the funds released in the first year are utilized, the funds needed for the second year should be released in one lot. Similarly funds for the third year

should be released as soon as funds released in the first year and 75% of second year are utilized. For utilization of funds, a certificate of the local authority should suffice.

8. For mobilization of financial resources for LCS, no new towns should be taken up for sewerage for the next three to five years. However, sewerage may be provided wherever it is found absolutely essential.
9. The budget provision should be separate water and for sanitation, which includes sewerage and also low-cost sanitation. Low-cost sanitation should be specifically mentioned separately under sanitation and appropriate provision be made. Low-cost sanitation should be included in the Minimum Needs Programme so that the provision cannot be diverted or reduced.
10. While legal provisions are necessary to support the programme, more important is that existing bucket or dry latrines in Central and State governments, semi-government, public undertakings, railways, defence and local authorities buildings should be first converted to water flush units; otherwise it may not be possible to enforce the law. The legal provisions can be challenged in a court of law. There are instances where private enterprises polluting the water course with factory effluent could not be prosecuted as the court upheld their views that local authorities discharging the raw sewage were of greater health hazards. The Employment of Manual Scavengers and the Construction of Dry Latrines (Prohibition) Act, 1993 can make the desired impact only when it is strictly enforced. Therefore, the State Governments may be persuaded and pressurized to understand the importance of enforcing the legal provisions contained in the Act.

11. For the use of households where construction of individual latrines is not feasible, community latrines at appropriate places in adequate numbers be constructed. Due to weak financial position of local authorities, it is not possible for them to construct community latrines from their resources. The State Governments are also not able to finance these as per requirement. Hence, it is suggested that 50% grant and matching loan at 4% rate of interest and 20 years repayment period be provided to local authority by the Government of India and the HUDCO.
12. The HUDCO allows only 10% of the estimated cost as implementation charges. This should be increased to 20% wherever needed.

Rural Sanitation

1. The existing Rural Sanitation Programme of the Govt. of India aims at assisting families below poverty line exclusively and giving them subsidy for construction of toilets. There is no programme for assisting other categories of families in the shape of subsidy who wish to go in for construction of toilets. The programme of Govt. of India, from the point of view of sanitation, is based on an artificial distinction, between families below poverty line and families belonging to other categories. It is not only the families below poverty line who do not have toilet facilities. The real position is that even most of the families above poverty line do not have toilet facilities. The Govt. of India programme, therefore, should be for all categories of people living in the rural areas. The possible help either in the form of subsidy or loan or both should be extended to all beneficiaries who want to have toilets in their houses irrespective of income groups. First come and first served should

be the criterion for the selection of the beneficiaries for providing toilets in their houses.

2. For making the rural sanitation programme a success, it is necessary that there is a provision for bank loans at differential rate of interest for those who are not in a position to take up construction of toilets on their own with available subsidy amount.

3. In the context of the size of the problem of sanitation in rural areas, as also the fact that there is constraint of resources, it is appropriate to discuss the technology, which is cost effective and which ensures wider coverage. The existing technologies of sewerage and septic tanks are simply out of question as their installation, operation and maintenance are cost prohibitive and resources are just not available for them. The technology should be cost-effective, affordable and ensuring wider coverage. The Sulabh's twin-pit pour-flush toilet technology is appropriate, affordable, eco-friendly, scavenging free and socio-culturally acceptable. This technology has been implemented on a large scale in our country successfully. 12 lac individual toilets have been constructed/converted by Sulabh.

4. Experience shows that it is not practicable to impose a uniform design of toilets on users from above. There should be several designs of pour-flush toilets meant for various income groups in rural areas. Sulabh has prepared 46 designs. There are as many as 24 designs with cost ranging from Rs. 500 to Rs. 50,000 which have been devised by Sulabh to suit people with different income levels. These designs can serve the purpose in different geographical and climatic conditions. Choice of designs should be left to beneficiaries who will take a decision keeping in view their requirements and the resources which they are able to mobilize.

5. Most of the houses in rural areas are very small, either mud built on semi-brick structure and there is not enough space for constructing toilets inside the houses. The programme of rural sanitation, therefore, will involve two elements: i) there should be a programme of construction of individual toilets where space is available and the beneficiaries have to be given financial support for construction of such toilets; ii) construction of public toilets in the rural areas, on account of constraint of space, is generally not favoured. Public toilets, however, should be constructed near Panchayat Bhawans, Village Markets, Health Centres, Bus Stands and other public places where people congregate in large numbers. Such public toilets should be constructed by NGOs and maintained by them on 'pay and use' basis.
6. The choice of agency for implementation of sanitation programme in rural areas is of considerable significance. Compared to urban situation where municipalities and municipal corporations have been in existence for quite some time and they have some experience of sanitation work in some form or other, in rural areas there is no such agency. Even Panchayats and Panchayat Samittees are not sensitized on the issue of sanitation. They have neither resources not the experience to take up sanitation work in rural areas on a large scale. Even logistically this becomes difficult because rural areas are scattered and far flung. The choice of agency, therefore, will be crucial to the success of rural sanitation programme. Since the work involved motivating, educating the people about evil effects of such toilets, these can be performed by the non-governmental organizations which have experience and infrastructure for taking up this work. It is in this context that the choice of NGOs is a must for the implementation of rural sanitation programme.

It will not be possible to make rural sanitation programme a success through Govt. machinery as Govt. functionaries working in rural areas have neither orientation nor experience of doing this type of job. It is, therefore, necessary that non-governmental organizations are involved in a big way for implementation of this programme. Such organizations, besides educating and motivating the people will also assist them in getting loan or subsidy for construction of toilets.

7. The entire range of activities starting with the information, education, communication (IEC) and also including implementation and follow-up should be given to the same NGOs in the respective areas. It has been experienced that if the work is divided among various organizations, it becomes a case of divided responsibilities which hampers programme of work.

8. The NGOs selected for implementing the sanitation programme should be allowed 15% of the estimated cost which would include start-up and IEC activities including implementation and follow-up services. In addition, 10% of the project cost should be allowed for training, support services, publicity which would include printing of booklets, literature, posters, organizing nukkad dramas etc. For setting up Sanitary Marts and Production Centres, subsidy as admissible under Govt. of India Sanitation Programme should also be paid to the NGO.

9. Capacity building is an important component of rural sanitation programme. Sulabh has prepared a four stage training model for rural development i.e. training at State level, District level, Block level and Panchayat level. Training of Trainers approach will be used at every level since they will be for carrying out training at next level below. At State level, two

trainers from each district will be trained in the implementation of the sanitation programme who will, in turn, train persons at District level drawn from Blocks so that they can take up construction of toilets in villages. While volunteers will act as motivators, the masons will take up construction of toilets in rural areas on behalf of willing households.

10. The school sanitation will form an important component of the programme as besides encouraging girls' education in the villages, it is through school teachers and students that the message of sanitary toilets will reach the villages. The NGOs which have been identified by the State Governments in a particular area for construction of toilets should also be allowed to do construction of toilets in schools in that area, as also impart training about maintenance and construction to teachers and students.

11. There should be total social mobilization for making the rural sanitation programme a success. Policy makers, village elders, school teachers, local leaders and prominent persons should be actively involved in the implementation of the sanitation programme.

12. The women in particular should be actively involved in the implementation of the programme right from the very beginning, as they are the worst sufferers on account of lack of toilet facilities. They will have dominant role in hygiene education. They are responsible for handling of drinking water, home sanitation and hygiene in their houses. It is, therefore, necessary that they are properly trained in these areas. The training programmes may be drawn-up in such a way that they attend the classes when they are free from the routine work.

13. It has been observed that by and large latrines built in rural areas specially in houses of people below the

poverty line, are not being used but the structure is being used for other purposes. Therefore, the implementation of the programme should be based on 'demand driven approach', and not as a traditional 'supply-driven subsidy oriented Government programme'.

14. The Grameen and nationalized banks should be directed to provide loan to beneficiaries for construction of water flush toilets in their houses on rate of interest as is being done for many other poverty alleviation programmes.

15. Proper delivery system is of prime importance to make the programme successful. The rural sanitation programme should not merely be hardcore approach i.e. construction of latrines but should promote a general awareness about health and personal hygiene. Hence NGOs are best suited to implement the programme as they are better equipped with the necessary infrastructure on software as well as hardware. But there should be proper screening of voluntary organizations with a view to assessing their capacity to undertake the programme with requisite experience in this field to undertake this programme. The responsibility for selecting the NGOs based on their experience, expertise and infrastructure available with them should rest with the District Magistrate or the State Government.

16. Creation of infrastructure at grass-root level by training two local youths and masons in each Panchayat in various components of water supply and sanitation such as construction of Sulabh Shauchalaya, sullage drains, pavement of roads, biogas operation and maintenance of handpumps, social forestry, smokeless chulhas etc. will accelerate the progress of rural sanitation programme considerably. It will also generate additional employment. At state, district

and block levels too, cadre of trained persons should be created to monitor the programme and guide the grass-root workers. The multiple type of work in which youths would be trained would enable them to earn their livelihood by getting modest amounts in return for their services rendered. They would act as a link between the implementing agency and the beneficiaries. The trained youths would inform the beneficiaries about the development programme contents and get the works done, either by themselves or by arranging other agency. They will also provide the follow-up service and educate the beneficiaries on use, operation and maintenance of toilets. This innovative approach developed by Sulabh can usher in a new era of improvement of sanitation and environment in the villages, besides generation of employment.

17. One of the reasons for poor progress of the sanitation programme is that the implementation charges given to voluntary organizations are completely unrealistic, which do not take into account fully the overhead expenses on the implementation of the programme. Under the existing arrangement, this is limited to 7.5% of the estimated cost with which the voluntary organizations are not able to ensure quality construction. This should be increased to 20%.

18. Community toilets should, as far as possible, not be provided in rural areas. Individual toilets should be preferred. However, public latrines may be provided in panchayat bhawans, schools, health centres, community centres, market places and bus stands etc.

19. The foremost work which requires to be done in rural areas is to make people aware of the environmental effects of dry latrines and matters pertaining to open defecation, home sanitation, food hygiene and personal hygiene. They have to be educated about

the advantages of sanitary toilets. Although there are various aspects of rural sanitation to be taken care of, there is no denying the fact that the main problem is that of providing proper latrines for the rural people.

20. The upgradation of the rural sanitation scenario will not improve simply because toilets are built. The programme should be accompanied by improvements in hygiene behavior. Hygiene improvements will result from the sustained practice of safe hygiene behavior, improved awareness and skills for a healthier environment in which people live. This can be achieved through a combination of convenient toilet facilities. But still hygiene improvement will be a critical factor. Hygiene awareness and changes in the public attitudes have, therefore, been recognized as essential elements in preventing the spread of diseases. The rural population needs to be helped in understanding sanitation if health considerations are to carry appropriate weight in their decision to build and maintain a latrine. The existing supply driven approach has proved to be a poor strategy. What is urgently required is the creation of right environment for sanitation promotion.

21. In regard to promotion of sanitation programmes, there are two views. The first view asserts that sanitation should be promoted as part of an integrated programme of rural development, or primary health care, or at least together with supply and hygiene education. Interventions combined in a single programme sometimes have a greater impact than they might, individually, and implementing them together can be cheaper. The most powerful argument for combining sanitation with other interventions such as rural water supply is that sanitation would otherwise be neglected. The

relatively high profile of water sector makes it a suitable vehicle to put sanitation into prominence and attract resources it needs. On the contrary, the second view asserts that sanitation does not always benefit from integration with other measures. The implementing agency may not have the appropriate staff or structure. Moreover, the decision to implement several interventions in a single programme does not guarantee that they will advance at the same pace. The pace of sanitation implementation is set not by administrative fiat but by consumer demand, so that it rarely matches the progress of other measures. In combined water and sanitation programmes, sanitation usually lags behind.

Notwithstanding these two divergent views, there is no general answer to the question whether sanitation should be promoted on its own, in free-standing sanitation programmes, or as part of integrated packages. The question should not be decided from theory or general principle, but on the merits of the case in each specific context. There may not even be a single answer for a whole country or region.

22. Since the problem of rural sanitation is enormous in size, a view has to be taken regarding the timeframe in which the problem has to be tackled and ultimately solved. The decision on timeframe is very important because that will determine the allocation of funds available in annual plans of the central and State Governments. On a realistic basis, it is suggested that a timeframe of 25 years should be kept in view and accordingly allocations should be made in the plans of Central and State government sectors.

Urban Sanitation

1. The foremost issue in urban sanitation relates to the adoption of technology in the context of constraint of resources earmarked for this programme. The

sewerage system is cost prohibitive as a result of which the coverage is sparse and partial. It will be advisable to cover a large number of towns and cities under the conversion programme in preference to laying of sewer in selected small number of towns. The technology of septic latrine is also not recommended as besides cost, scavengers will be required to clean them which goes against the very concept of liberation of scavengers. The only situation in the circumstances is the two-pit pour flush technology which is affordable and requires least quantity of water for flushing.

2. The programme of low-cost sanitation and liberation of scavengers is being implemented since 1980-81. We have now considerable experience of this programme. The pattern of financial assistance to the beneficiaries is crucial to the programme. When this programme was initiated there was a provision for 50% subsidy and the balance 50% as loan to the beneficiaries. Till 1988-89, the programme went on smoothly and achieved reasonably good progress commensurate with the availability of funds. Subsequently, there was change in the financing pattern used on income level of beneficiaries. The beneficiaries have been put in four categories namely those belonging to economically weaker sections, low income group, middle income and high income groups. Our experience of change in the pattern of financial assistance to the beneficiaries is that it has become cumbersome. The correct and precise categorization in four groups is rather difficult and by and large the overwhelming majority of the people in urban areas are financially weak. The change in procedure has rather slowed down the pace of implementation of the programme. It will not be possible for the Government of India to achieve the target of making the country scavenging free by 2015

unless the financial pattern of 50% subsidy and 50% loan, which existed prior to 1988-89, is restored.

3. Another point relates to the release of funds to the State Government by the HUDCO. There are a number of instances in which the funds released by HUDCO to the State Governments have not been passed to the local bodies in time and there has been hold up at State Government level for considerable time with a view to improving ways and means position of the State Governments. It is, therefore, suggested that on the pattern of arrangements made by CAPART, HUDCO should release funds direct to NGOs so that there is no delay in utilization of funds earmarked for this programme. There should be no problem in releasing funds direct to the NGOs as the State Governments will be required to recommend only those NGOs which have commitment and dedication to the programme and have also requisite infrastructure. If such NGOs are not available the funds should be placed at the disposal of Local Bodies with a direction that the same should be drawn duly counter-signed by the Collector of the district and kept in the Banks instead of treasury so that it can be easily withdrawn as and when required. The advantage in keeping the fund in Banks is that it will be utilized only for the purpose, for which it is meant.

4. It has been our experience that the new toilets which have been constructed without enclosures are sometimes not used by the beneficiaries. The enclosures have not been put up as they are not part of the cost of toilets. It is, therefore, suggested that in the case of new constructions, there should be provision for enclosures and this should be included in the cost of the toilets.

5. Community toilets are needed for slum and pavement dwellers, rickshaw pullers and floating

population. The financial position of local authorities is not such that they can meet the cost from their own resources. The Government of India/HUDCO should provide financial assistance for the construction of these toilets on the same financing pattern as applicable to the provision of individual toilets to economically weaker sections.

6. Slum and pavement dwellers are worst suffers in absence of toilet facilities. Hence, the provision of sanitary facilities, especially the access to toilet facility to slum and pavement dwellers should be given priority to improve their health and living environment. Attempts should be made to provide public toilets where necessary and feasible to discourage open defecation in and around slum areas' poverty stricken settlements. Accessible and affordable sanitation services are crucial to improving the lives of slum dwellers.

7. The community toilets are mainly operated by the municipal authorities, but these remain so insanitary that people prefer to go for open air defecation rather than using them. Hence, in order that the community toilets could be used, it should be made obligatory for the local authorities to hand over their operation and maintenance to the NGOs having experience in this field and these should be operated on 'pay and use' basis.

8. The work of conversion/construction of sanitary toilets should be awarded to select NGOs who have got experience, manpower, infrastructure and credibility. The Government should prepare a list of NGOs who fulfil these criteria.

9. The National Commission on Population constituted by the Planning Commission after a great deal of deliberations has recognized the role of non-governmental organizations in the implementation of urban sanitation programmes: It said:

(i) Information, education and communication (IEC) constitute an important element in the implementation of sanitation programme, particularly the conversion of dry latrines into low-cost sanitary toilets.

(ii) Voluntary organizations should be engaged right from the beginning till the end i.e., motivation, implementation and follow-up as it would be desirable to give them at least 15% as implementation charges.

(iii) IEC implementation and follow-up may be entrusted to single agency.

(iv) The Government should be the facilitator rather than executor in sanitation programmes. The role of the Government should be limited to resource mobilization, monitoring and evaluation of the programme.

The promotion of sanitation generally requires a cadre of well-trained people in the field. They should have the same cultural roots as those they serve, and they must understand the technical aspects of low-cost sanitation. Above all they must be sensitive to the perceptions of the communities and able to win their confidence, deal with their doubts, and stir their enthusiasm. To recruit, train, establish and manage such a cadre of field workers is a far greater challenge than simply to manufacture and sell latrine components. Such individuals seem to be found more often in voluntary organizations than in government service, and more sanitation initiatives are probably undertaken by non-governmental organizations than by governments.

Training and Rehabilitation of Liberated Scavengers

1. A central committee at the Government of India level should be formed under the chairmanship of the Deputy Chairman of the Planning Commission for reviewing the programme of training and

rehabilitation of liberated scavengers in the entire country.

2. Similarly a Steering Committee under the chairmanship of the Development Commissioner may be set up at the State level to lay down policies and guidelines of training and rehabilitation programmes and issue directions where found necessary.

3. Suitable vocations in which training needs to be imparted have to be properly identified, keeping in view their marketability and norms of job requirements.

4. The work of providing training to the liberated scavengers and their dependents may be entrusted to reputed non-official organisations possessing requisite expertise, resources and experience in the field.

5. The work of training and rehabilitation of scavengers should be entrusted to the same organization and should not be given to different agencies. This will ensure an integrated approach to the problem and avoid gaps in rehabilitation.

6. The State Government should make available land, building for training, hostel for trainees, necessary tools and equipments with all other required accessories and infrastructure to the agency made responsible for training and rehabilitation of the liberated scavengers and their wards or may give a grant to the agency made responsible for training. Till such time this is not made available, a suitable building may be hired.

7. Regional training institutes should be set up for imparting skills to the liberated scavengers and their dependents in various trades and vocations.

8. The certificates given to the trainees by the agency imparting training should be recognized by the

Central as well as the State Governments, so that the trainees may not face problems at the time of seeking jobs.

9. The liberated scavengers and their wards should be trained as far as possible in vocations suitable to their aptitude. However, the main thrust on selecting vocations for training should be the employment potential and demand so that after training, the problem of getting employment is not faced.

10. The scavengers and their children being underprivileged, their mental aptitude and ability cannot be matched with average students. Hence, the duration of training should be kept sufficiently long so that the trainees may develop efficiency in the skills that may enable them to compete in the job market.

11. Since the liberated scavengers lose their source of income, they should be provided adequate stipend during the training period, so that they may continue and complete the training. It is suggested that the amount of stipend should be commensurate with the current cost of living.

12. For training, funds should be made available by the State Government to the organization made responsible for training on the basis of fee per trainee. The amount of fee should be adequate to cover all the training costs.

13. It has been our experience that the liberated scavengers after training do not get employment as in the absence of proficiency, they are unable to compete with other candidates. Hence, it is necessary that after completing the training, they are kept as apprentices in the government, semi-government, public undertakings, non-governmental organizations (NGOs) and commercial firms for a year. During this period, the Government should pay a

stipend of Rs. 1,000 per person per month to meet the boarding and lodging expenses. This procedure will also enable the organizations to assess the performance of the candidates and will facilitate their employment in the same organizations where they are working as apprentices.

14. After training, the trainees should be helped in getting employment in government and non-government organizations, public and private undertakings; and assisted in securing financial assistance for self-employment. A person in the training institute has to be made responsible solely for providing this assistance to the trainees.
15. The amount of subsidy and loan given to the liberated trained scavengers for self-employment should be adequate enough to meet the full requirement of the business or manufacturing units they want to set up. The financial assistance should also include margin money needed for the trade.
16. The trained liberated scavengers should be helped in forming cooperatives for setting up business/trade and also for marketing the products. The government, semi-government organizations and public undertaking should be directed to give priority to the purchase of goods manufactured or sold by the trained scavengers.

Suggestions for the Government of India

1. The Prime Minister should convene a special conference of the Chief Ministers to discuss the issue of liberation of scavengers. The conference should have the following major points on its agenda:

   (a) To know the viewpoints of the Chief Ministers on sanitation especially the liberation of scavengers.

(b) To find out the resource constraints for the scavengers' liberation programme.

(c) To know their opinion on fixing a time-limit to end scavenging in the States and Union Territories.

(d) To evolve a common strategy for the implementation of the programme including the adoption of an affordable and cost-effective sanitation technology.

(e) To discuss the simplification of funds transfer process in order to ensure the timely utilization of money to achieve the target within a broad timeframe.

(f) And other related matters.

2. Substantial increase in the quantum of funds for the sanitation programme with a view to liberating and rehabilitating all scavengers by 2010.

3. The work of liberation and rehabilitation of scavengers should be integrated at the local, State and Central levels and an appropriate mechanism be accordingly evolved. This will avoid duplication of work and the consequent delay.

4. While the Central and State Government have undertaken several target-oriented programmes for the abolition of scavenging, the experience, however, suggests that the State Governments do not have the kind of commitment which the task deserves and demands. It is, therefore, suggested that the State Government should be persuaded to take up the programme in right earnest, indicating their full commitment to solving the problem of scavenging.

5. The sanitation programme should be treated as part of the Central Sanitation and Health Care Plan for which there should be a separate budget allocation.

6. In the absence of reliable data on the total number of scavengers and their families, total number of liberated scavengers and those rehabilitated in different occupations, total number of dry or bucket latrines and total number of conversions, etc., the magnitude of the problem cannot be assessed nor appropriate measures suggested to fully meet the situation. It is, therefore, suggested that the Central Government may like to undertake benchmark surveys to collect such data including socio-economic surveys to fully know the working and living conditions of the scavengers, and their problems – social, economic and occupational.

   In order to do this the concerned Ministries/ Departments of the Central, State Governments and local bodies should sponsor research projects on various aspects of the problem to be conducted by universities/research/training institutions and voluntary organizations. The research projects so sponsored should be exploratory, descriptive, diagnostic and evaluative in nature.

7. The Centrally-sponsored scheme of pre-matric scholarships for the children of those engaged in unclean occupations should be implemented through reputed non-governmental organizations. The scheme should provide the opening of residential schools for the children of scavengers studying up to the 10th standard.

8. It has been observed that the families of scavengers mostly live in slums, near the dumping ground of the refuse, which are most insanitary and unhygienic without the basic amenities like drinking water, electricity, sanitation, pucca streets, etc. It is therefore, necessary that the Government should acquire suitable land and allot house sites measuring about 125 sq.m. to each scavenger family or construct

and allot houses to them. A similar Centrally-sponsored scheme was in existence during the Fourth Five year Plan which may be revived.

9. The local authorities and NGOs which have done good work in the field of liberation of scavengers should be given suitable awards at annual functions.

Suggestions for State Governments

1. The State Government should also conduct State-level surveys to make overall assessment of the magnitude of the problem of scavenging. Such surveys should contain the following information:

   a) Total number of liberated and unliberated male and female scavengers.

   b) Total number of scavengers (both male and female) engaged in scavenging.

   c) A detailed demographic, and socio-economic-cum-educational profile of the scavenger families.

   d) Total number of boys and girls in each scavenger family in need of vocational training.

   e) Total number of bucket or dry latrines so far converted into water-borne pour-flush latrines.

   f) Total number of bucket or dry latrines still in existence.

2. Make detailed assessment of the total quantum of funds required for:

a) Conversion of bucket or dry latrines; and

b) Liberation and rehabilitation of scavengers by the stipulated time period.

3. Quick and timely distribution of funds earmarked for the municipalities and municipal corporations,

other agencies and organizations involved in the implementation of the programme for the liberation and rehabilitation of scavengers.

4. The progress of the programme for the liberation and rehabilitation of scavengers should be periodically reviewed and appropriate remedial measures on the basis of the received feedback be taken.
5. Allocate at least one per cent of their total budget on sanitation to boost the programme for the liberation and rehabilitation of scavengers and complete the task during the specified time-schedule.
6. Make a time-bound programme, after having made an assessment of the extent of work to be done and the resources available for the purpose.
7. Encourage non-governmental organizations, especially those having expertise, infrastructure and credibility, to help the State Government in completing the task within a reasonable timeframe.
8. Involve grass-root non-governmental organizations in its promotion of a campaign for the eradication of scavenging and the training and rehabilitation of scavengers.
9. There is a lack of adequate sanitation facilities in crowded places in the cities and towns like railway stations, bus stops, cinema houses, hospitals and important business centres etc. Community latrines should, therefore, be constructed at such places and their operation and maintenance should be entrusted to the reputed NGOs having a long and wide experience in this field.

Suggestions for local Authorities

1. Each local authority should undertake a survey at its

level to have with it a reliable data-base for the implementation and execution of the programme for abolition of scavenging and for the liberation and rehabilitation of scavengers. The data should contain information on: (a) the number of scavengers (both male and female) working in private houses, (c) a detailed demographic, socioeconomic and educational profile of the sons/daughters of scavengers in need of training and rehabilitation.

2. The local authorities should ensure that the money allocated for sanitation is not diverted to other projects.
3. The local authorities should allocate a certain percentage of their total budget for the liberation of scavengers.
4. The local authorities should work to ensure the total elimination of scavenging within a broad time-frame.
5. The local authorities should seek the involvement of reputed non-government organizations in the project formulation and programme implementation and thus expand the sanitation movement at the town level.
6. The local authorities should also join hands with non-governmental organizations in promoting sanitation awareness programmes through campaigns, media, publicity and education.
7. The local authorities should, as far as possible, absorb the liberated scavengers in other jobs.
8. The local authorities should make efforts in organizing/promoting training programmes for scavengers and their dependents.
9. The local authorities should organize/arrange for the health, housing, education and other welfare services for the families of both liberated and unliberated scavengers either through their own

resources or with the help of other local/State level departments/bodies. The provision of such services would ensure the meeting of basic minimum needs of the scavengers.

Suggestions for Non-Governmental Organisations

The liberation and rehabilitation of scavengers cannot be accomplished by the Government alone. It also calls for the involvement of non-governmental organizations (NGOs) engaged in various social welfare activities, especially those relating to the welfare of the most disadvantaged people.

It is true that only a few voluntary organizations have chosen to work exclusively for the cause of scavengers, but the work done by the Harijan Sevak Sangh, Gandhi Smarak Nidhi, Safai Vidyalaya, Rajasthan Institute of Local Self-Government, Institute of Social Studies Trust and Sulabh International deserve special mention.

Bearing this in mind, the author would like to offer the following suggestions:

1. Selection of such pockets or areas for all-round development work in all cities and towns where residents predominantly belong to scavenger castes or other backward communities.
2. Identification of scavenger families whom the concerned organization would like to help in respect of their health, hygiene, education, training, rehabilitation and social upgradation.
3. Frequent and regular visits by the functionaries of the voluntary organizations to the scavenger families and to the areas predominantly inhabited by people belonging to scavenging castes/communities in order to create rapport with them, eventually leading to winning their confidence in the functioning of the voluntary organizations.
4. Carrying out campaigns at the local, State and

national levels to generate a social climate conducive to the satisfactory resolution of the problem of scavengers (and their families) in regard to their socio-economic development, their liberation from the demeaning and dehumanizing occupation, their training, their job placement and their social assimilation into the mainstream of the society.

5. Organising special functions/gatherings/marches/meetings on October 2 every year to highlight the problems and predicaments of scavengers and thereby create a favourable public opinion and social and political sensitivity and commitment towards completing the task of liberation and rehabilitation of scavengers.
6. Taking up a supporting role for and on behalf of the liberated and unliberated scavengers and working with them in self-help endeavours.
7. Setting up scholarships for meritorious students of the scavenger families and contributing towards their educational advancement.
8. Approaching media (Press, Doordarshan, Television and Akashvani) to highlight the problems of scavengers as well as progress of the work done for their liberation and rehabilitation both by the governmental and non-governmental organizations.
9. Organising seminars and conferences to discuss liberation and rehabilitation of scavengers.
10. Publishing booklets, brochures, posters, etc., to create wider public awareness about the problems of scavengers and the ways to deal with them.

Suggestions for Action Sociologists

1. Action sociologists must pledge to engage themselves in constructive social action at both macro and micro levels.

2. They should decide for their individual and collective action/intervention in such critical problems and issues of wider social concern as relate to the life of the poor, socially backward and politically disadvantaged people. The liberation of scavengers can legitimately be one such area.
3. Action sociologists should define their role and the parameters of their action and involvement and devise strategies/approaches to solve the problems faced by scavengers and their family members.
4. In conformity with the commitment to the welfare, development and empowerment of the weaker and vulnerable sections of society, the action sociologists have to produce serious writings on various facets of scavengers' problems. As a necessary precondition in the discharge of this much needed academic function, the action sociologists will have to do the following:

    a) Preparation of data-base on scavengers and scavenging by themselves or through their research students, along with conducting problem-focused or issue-based research studies (surveys) in their respective cities and towns all over the country.

    b) Writing books, articles in different languages with a view to highlighting the problem of scavenging in the country. These books and articles should reflect social concern for the redemption and rehabilitation of scavengers.

    c) Encouraging Ph.D. Students in their respective departments/universities/institutes to select such topics for their dissertations as are directly related to scavenging.

    d) Organize conferences/seminars/workshops in

their respective cities and towns on such themes as may have a direct bearing on scavenging.

e) Approaching the media (Press, Doordarshan, Television and Akashvani) to disseminate their views to a wider audience to sensitize people on the problems and predicaments of scavengers.

5. Action sociologists should undertake interventionist action in the following manner:

a) Selecting groups of students and preparing them to do voluntary work in such localities as are inhabited predominantly by scavengers.

b) Asking the NSS (National Social Service) Scheme units in their colleges and universities to organize special educational and awareness campaigns in the areas inhabited largely by scavengers or other lower castes.

c) Arranging for healthcare, educational programmes, clean water, waste disposal, recreational and environmental upgradation facilities, etc. in those pockets/areas where scavengers of their city/town reside. This can be done by approaching senior officials of the concerned local government and non-governmental organizations (NGOs).

d) Associating themselves with the programmes and activities of such local voluntary action groups as are already engaged or are willing to be engaged in liberating and rehabilitating the scavengers.

BIBLIOGRAPHY

Ahmed, Z. (1980) Evaluation of Low-Cost Water-Seal Latrine in Bihar (Mimeograph), Department of Sociology, Patna University, Patna.

Ambedkar, B.R. (1936) Annihilation of Caste with a Reply to Mahatma Gandhi, Bhushan Press, Mumbai.

—— *(1943)* Mr. Gandhi and the Emancipation of the untouchables, Thacker, Mumbai.

—— *(1951)* What Congress and Gandhi Have Done to the Untouchables, Thacker, Mumbai.

—— *(1946)* Who were the Shudras? How they came to be the 4th Varna in the Indo-Aryan Society, Thacker, Mumbai.

—— *(1948)* The Untouchables : Who were they and why they became Untouchables, Amrit Book, New Delhi.

—— *(1969)* The Untouchables, Bharatiya Baudha Sikhsha Parishad, Balrampur (U.P).

Bailey, F.G. (1958) Caste and Economic Frontiers, Oxford University Press, London.

Barve, B.N. (1958) The Scavengers' Living Conditions Enquiry Committee, Government of Maharashtra, Mumbai.

Beteille, Andre (1969) Social Inequality, Penguin.

—— (1969) Caste : Old and New, Asia Publishing House, Mumbai.

Bose, N.K. (1975) The Structure of Hindu Society, Orient Longman, New Delhi.

Black, M. (2001) Conference Report – First International Conference on Ecological Sanitation, 5-8, November 2001, Nanning, China

Barret, M. (2001) Ground Water and Sanitation, First International Conference on Ecological Sanitation, 5-8 November, Nanning, China.

Brown, L. R. (1998) State of the World 1998, Earthscan Publications, Lita London.

Calvert, P. (1998) A Positive Experience with Composting Toilets in India – Kerala Case Study, Paper Presented at the Centre for Science and Environment Conference on Health and Environment, July, 1998, New Delhi.

Cairncross, Sandy (1992) Sanitation and Water Supply : Practical Lessons from the Decade, Water and Sanitation Discussion Paper Series, The World Bank, Washington.

Chatterjee, Mary (1975) Conjugal Roles and Social Organisation in an Urban Sweeper Community, Unpublished Ph.D. Thesis of B.H.U.

Chaudhary, Partima (1988) "Problems of Female Scavengers of Patna", Changing Status of Depressed Castes in Contemporary India, (Ed. S.N. Choudhary), Delhi

Chaudhary, S. N. (2001), Occupatinally Mobile Scavengers Har-Anand, Delhi.

Chintamani, C.Y. (1901) Indian Social Reform, Thomson & Co., (ed.), Chennai.

Dass J.R.J. (1968) Bhangi Ksht Mukti and Bhangi Mukti, Harijan Sewak Sangh, Delhi.

Das, B. (1996) "Dr. Baba Saheb Ambedkar Aur Bhangi Jatiyan, Dalit Today, Lucknow.

Eisenstadt, S. N. (1965) Modernization, Protest and Change, Prentice Hall, New Jersey.

Feachem, R.G. et al. (1983) Sanitation and Disease : Health Aspect of Excreta and Waste Management, John Wiley and Sons, Chichester.

Fuchs, Stephen (1981) At the Bottom of Society : The Harijan and other Lower Castes, Munshi Ram Manohar Lal Publications, New Delhi

Gandhi, M.K. (1954) The Removal of Untouchability, compiled by B. Kumarappa, Najjiram Publishing House, Ahmadabad.

Government of India, Planning Commission (2002) National Human Development Report 2001, Oxford University Press, New Delhi.

—— *UNICEF (1978) National Seminar on Low-Cost Techniques for Disposal of Human Wastes in Urban Communities,* Kolkata.

—— *(1982) UNDP Regional Conference on Low-Cost Pour-Flush Latrine, Udaipur.*

—— *(1982) UNDP Regional Seminar on Low Cost Sanitation, Ministry of Works and Housing, New Delhi.*

—— *(1961) UNDP Regional Seminar on Social Mobility Among the Sweepers of India, Ministry of Home Affairs, New Delhi.*

—— *(1961) UNDP Report in Customers Rights and Living and Working Conditions of Scavengers in Two Towns, Census of India Monograph Series, Vol. I, Part XI-D, Ministry of Home Affairs.*

—— *1992, Proceedings of the National Seminar on Rural Sanitation, Ministry of Rural Development, New Delhi.*

Ghurye, G.S. (1957) Caste and Class in India, Asia Publishing House, Mumbai.

—— Caste class and occupation, Popular Prakashan, Mumbai.

Hardoy Jorge E. Mitlin Diana and Satterthwaita David, (2001) Environmental Problems in Urbanising World, Earthscan Publications, London.

Hutton, J. (1961) Caste in India : Its Nature, Function and Origin, Oxford University Press, Oxford.

Issacs, Harold R. (1965), India's Untouchables, Asia Publishing House, Mumbai.

Jayakar, Samuel (1987), Rural Sweepers in the City : A description of the Economic, Social and Religious Factors in their Changing Culture, The Wesley Press, Mysore.

Jessop, Bob (1972,) Social Order, Reform and Revolution, Macmillan, London.

Karlekar, Malavika (1979), Poverty and Women's Work : A Study of Sweeper Women in Delhi, Shakti Books, New Delhi.

Lynch, O.M. (1969), The Politics of Untouchability, Columbia University Press

Mahant, N.M. (1949), Mehtar Mimansa, Harijan Sewak Sangh, Bikaner.

Mahar, Michael J. (1973), Untouchables in Contemporary India, The University of Arizona Press, Tuscon.

Malkani, N.R. (1966), The Report of Committee on the Customary Rights to Scavenging, Ministry of Home Affairs, Govt. of India, New Delhi.

—— Report cf Scavenging Conditions – Enquiry Committee, Ministry of Home Affairs, Central Advisory Board for Harijan Welfare.

—— (1969) Clean People and Unclean Country, Delhi : Harijan Sewak Sangh.

Menon, A.S. (1971), Wage Board for Municipal Workers, Government of India.

Mulkraj Anand (1956), Untouchables, Jaico Publishing House, Mumbai.

Morgan, P. (1999), Ecological Sanitation in Zimbabwe : A Compilation of Manuals and Experiences Conlon Printers, Harare, Zimbabwe

Nagar, Amrit Lal (1980), Nachyo Bahat Gopal, Rajpal & Sons, Delhi.

National Sample Survey Organisation (1999), Drinking Water, Sanitation and Hygiene in India, 54th Round (January – June 1998) Report 449, Department of Statistics, Delhi.

Natrajan, S. (1959), A Century of Social Reform in India, Asia Publishing House, Mumbai.

Oommen, T. K. (1990), Pretest and Change, Sage Publications, New Delhi.

Pathak, Bindeshwar (1991), Road to Freedom – A Sociological Study on the Abolition of Scavenging in India, Motilal Banarsi Dass Publishers, Delhi.

—— (1981) Sulabh Shauchalaya : A Simple Idea That Worked, Amola Prakshan, Patna.

—— (1982) Sulabh Shauchalaya : A Study of Directed Change, Amola Press and Publications, Patna.

—— (1992) Action Sociology and Development, Concept Publishers, New Delhi.

Patil, Sharad (1982) Dasa-Sudra Slavery : Studies in the Origin of Indian Slavery, Allied Publishers, New Delhi.

Ram, Jagjivan (1980) Caste Challenge in India, Vision Books Pvt. Ltd., New Delhi.

Ratan, Ram (1960) "The Changing Religion of Bhangis of Delhi : A Case of Sanskritization" in L.P. Vidyarthi (ed.) Aspects of Religion in Indian Society, Meerut, Ramnath, Kedarnath.

Ribeiro, Edgar F. (1985) Improved Sanitation and Environmental Health Conditions – An Evaluation of Sulabh International's Low Cost Sanitation Project in Bihar, Ministry of Urban Development, Govt. of India

Roy, A. K. (1984) Manual on the Design, Construction and Maintenance of Low-Cost Pour-Flush Water-Seal Latrines in India, TAG Technical Note No. 10, UNDP International Project, The World Bank, Washington.

Roy, Burman B. K. (1968) : Social Mobility Among the Sweepers of India

Simpson – Hebert, M. (2001) "Ecological Sanitation and Urban Sustainability". First International Conference on Ecological Sanitation, 5-8 November Nanning, China.

Schonning, C. and Stenstron, T.A. (2004) Guidelines for the safe use of Urine and Faeces in Ecological Sanitation Systems, EcoSanres. SEI, Stockholm, Sweden.

SIDA (Swedish International Development Agency) Ecological Alternatives in Sanitation, Stockholm, Sweden.

Sagar, S.L. (1975) Hindu Culture and Caste System in India, Uppal Publications, Delhi.

Sharma, R.S. (1950) Sudras in Ancient India, Motilal Banarsidass, Delhi.

Salappa, IPD (1976) Committee on Improvement of Living and Working Conditions of Sweepers and Scavengers, Government of Karnataka

Shyamlal (1992) The Bhangi : A Sweeper Caste – Its Socio-Economic Portraits, Popular Prakashan Mumbai.

—— (1970) "Educational Development Among the Bhangis in Jodhpur", Vol. 7, No. 1

—— (1972) "Jodhpur Ke Bhangi", Harijan Sewa, Vol. 21, No. 1

—— (1972) "A Scheme for Bhangi's Welfare in Jodhpur", Harijan Sewa, Vol. 21, No. 6.

—— (1973) "Sanskritization and Social Change Among the Bhangis of Jodhpur : A Case Study", Indian Journal of Social Work, Vol. 34, No. 1

—— (1975) "Caste and Occupational Mobility in Jodhpur Bhangi Bustee", Journal of Social Research Vol. 18, No. 2

Sinha, Bakshi D. and Arun k. Ghosh (1991) Sanitation and Environment Evaluation of Low Cost Sanitation in Gujarat, Arnold Publishers, New Delhi.

—— (1990) Evaluation of Low-Cost Sanitation : Liberation, Training and Rehabilitation of Scavengers, Arnold Publishers, New Delhi.

—— (1994) Restoration of Human Dignity – Impact of Development Programme on Scavengers, Arnold Publishers, New Delhi.

Singh, Yogendra (1973) Modernization of Indian Tradition, Thomas Press, New Delhi.

—— (1980) Social Stratification and Change in India, Manohar Publishing, New Delhi.

Singh, Ram Gopal and Gadkar, R.D. (eds.) (2004) Restoration of Human Rights and Dignity to Dalits, Manak Publications, New Delhi.

Singh, Vinod (1991) Social Structure of Bhangi Community, A Sociological Study of KAVAL Towns of U.P., BHU, Varanasi.

Singer, Milton B. (1959) Traditional India : Structure and Change, American Folklore Society, (ed.), Philadelphia.

Srinivas, M.N. (1966) Social Change in Modern India, Berkeley, University of California Press

—— (1962) Caste in Modern India, Popular Prakashan, Mumbai.

Srivastava, B.N. (1992) "State Action for the Liberation of Scavengers", in Report of the National Seminar on Action Sociology and Liberation of Scavengers, Feb. 8-10, New Delhi.

UN (United Nations) (2000), Millennium Declaration, New York.

—— (2000) "Progress Made in Providing Safe Water Supply and Sanitation for All during the 1990s" Report of the Secretary General, Commission on Sustainable Development, New York.

—— (2002) "Millennium Indicators : India", Statistics Division, New York.

—— (2002) Report of the World Summit on Sustainable Development, 26 August, Johannesburg.

UNCHS (United Nations Centre for Human Settlements) (1996), An Urbanizing World : Global Report on Human Settlements, Oxford University Press, Oxford.

UNDP (United Nations Development Programme) 2003, Human Development Report.

UNEP (United Nations Environment Programme) 2003, Global Environment Outlook, London.

UN-HABITAT (United Nations Human Settlements Programme) 2002, "Report of the Expert Group Meeting on Urban Indicators" Nairobi.

Wagner, E.G. and Lanoix, J.N. (1958) Excreta Disposal for Rural Areas and Small Communities, World Health Organisation, Geneva.

Winbald, U. and Kilama, W. (1985) Sanitation without water, Macmillan, London.

Winbald, U. and Simpson – Hebert, M. (2004) Ecological Sanitation, Stockholm Environment Institute, Stockholm, Sweden.

WHO (World Health Organisation) (1992) Improving Water and Sanitation Hygiene Behaviour for the Reduction of Diarrhoeal Diseases, Geneva.

Winbald, Uno and Mayling Simpson Hebert (eds.) (2004), Ecological Sanitation, Stockholm Environment Institute.

WEHAB Working Group (2002), "A Framework for Action on Water and Sanitation", Paper presented at the World Summit on Sustainable Development, 26 August, Johannesburg.

WHO (World Health Organisation) (1997), Health and Environment in Sustainable Development : Five Year After the Earth Summit, Geneva.

—— (2003), "Major Causes of Death Among Children under Five, Worldwide 2000", Overview on Child Health.

—— (2000) Global Water Supply and Sanitation Assessment 2000, Geneva.

World Bank (2004), World Development Report 2004, Washington D.C.

WSSCC (Water Supply and Sanitation Collaborative Council) (2000) "Wash Facts and Figures", Geneva.

Yashwant Raj (1997) The Ugly Face of Manual Scavenging in India. An unpublished Dissertation submitted at the Indian Institute of Public Administration, New Delhi.

Books on Hinduism and its Sources :

Apad-Dharma in the Mahabharata : How to face Calamities at Personal, State and Global Levels—SATYA P. AGARWAL

Candipathah : Incorporating *Sridurgasaptasati (Devimahatmyam)* and the associate Hymns (with text in *Devanagari* and Roman— Tr. & Introductory Essay by ALLAHABADIA PRAN NATH PANKAJ, in association with BHAVANA PANKAJ

Essays on Hinduism— KARAN SINGH

Gayatri : The Profound Prayer— R.K. MADHUKAR

Hindu Samskaras—RAJBALI PANDEY

Hinduism : Scriptures & Practices—PRABHA DUNEJA

Puja and Samskara— MUSASHI TACHIKAWA, SHOUN HINO & LALITA DEODHAR

Ramayana : The Ancient Indian Epic by Sage Valmiki describing the experiences, achievements and Journey through the life of Lord Rama)— RAVINDRA SHEKHAR SHUKLA

Ramayana at a Glance—SADGURU SANT KESHAVADAS

Shri Ramacharitamanasa or the Holy Lake of the Acts of Rama: Basic text in Devanagari with Hindi & Eng. Tr.—R.C. PRASAD Enlarged Ed. Compact Ed. Romanized Ed.

The Son of the God of Wind : The Exploits of Sri Hanumana—ATMA RAM SHARMA

The Sraddha : The Hindu Book of the Dead, A treatise on the Sraddha Ceremonies—R.C. PRASAD

The Upanayana : The Hindu Ceremonies of the Sacred Thread — Tr. R.C. PRASAD

The Vivaha : The Hindu Marriage Samskaras—BHAIYARAM SARMA, Tr. R.C. PRASAD